THE THEORY OF ENVIRONMENT

PART I

An Outline of the History of the Idea of Milieu, and its Present Status

BY

ARMIN HAJMAN KOLLER, PH.D.

Instructor in German

The University of Illinois

The Collegiate Press

GEORGE BANTA PUBLISHING COMPANY

MENASHA, WISCONSIN

1918

Copyright, 1918

By Armin H. Koller

TO

MY PARENTS

Contributions by

Jeff Steinport

ISBN 978-1977638267

PREFACE i

INTRODUCTORY REMARK Meanings of the Word "Milieu" iii

I A Sketch of the History of the Idea of Milieu Down to the Nineteenth Century 1

II A Sketch of the History of the Idea of Milieu Since the Beginning of the Nineteenth Century 11

    Anthropo-geography, Geography and History 11

    Geography and History 18

    More Recent Anthropo-geographical Treatises 28

    Primitive Peoples and Environment 30

    Society and Physical Milieu 31

    Government, War, Progress, and Climate 32

    Climate and Man's Characteristics 34

    Man's Intellect and Physical Environment 35

    Religion and Physical Milieu 36

    Climate and Conduct 37

    Climatic Control of Food and Drink 39

SUMMARY 41

APPENDIX 45

PREFACE

In 1912 (see *Publications of the Modern Language Association of America*, Vol. 28, N. S., Vol. 21, 1913, Proceedings for 1912, p. xxxix), I called attention to the Herder-Taine problem on milieu. The paper discussing that problem awaits the completion of another paper entitled "Herder's Conception of Milieu." The latter was my starting point. Setting about to inform myself on the history of the theory, I determined to obtain for myself, if possible, a tolerably complete idea, at least in its essentials, of the theory of milieu, to see where the theory led to, where it started from, what changes it has undergone, and what were its ramifications. My plan was to state briefly my findings in a chapter preparatory to stating Herder's idea of milieu. As guide-posts were lacking, at least I knew of none, I was bound to seek by accident and for a number of years. In stumbling along, I first chanced upon the Herder-Taine problem. When my material swelled to proportions that could not be controlled in part of a chapter or in a chapter, I had to separate it, by its main divisions, into parts. The question arose, should it be a *concrete* treatise on environment. I soon found that to be, at least for the time being, beyond my province and also beyond my present purpose; besides, it would have swerved me too far afield; moreover, it would have had to be limited to a small portion of the subject. My present concern in this theory being genetic and historical, it seemed best to assemble all the sources one could find bearing on the history of the theory and to indicate the trend of its development in a rough preliminary sketch. Such a sketch is a requisite first step and perhaps a modest contribution to a history of the theory under consideration. The first part of this sketch is herein given. The original plan, mentioned above, of a prefatory chapter to Herder accounts for the retention of untranslated passages in the text of this part, a practice to be eschewed in the subsequent parts of this study which are to appear shortly.

Nearly all the material was collected by October, 1915, and this manuscript was finished early in January, 1917.

I gratefully acknowledge my indebtedness to Professor Martin Schütze of the University of Chicago for the suggestion, made in 1907, to find out what Herder's idea of milieu is; to my friend and former colleague at the University of Illinois, Dr. Charles C. Adams (now Assistant Professor of Ecology at Syracuse University) for references given me at my request (but he is in no wise to be held responsible for the bringing in of these references); and to my good friend and colleague, Professor John Driscoll Fitz-Gerald of the University of Illinois for a number of helpful suggestions given when reading the manuscript and for assisting with the reading of the galley proof.

Armin H. Koller.

Champaign, Illinois,

April, 1918.

INTRODUCTORY REMARK
Meanings of the Word "Milieu"

Before entering upon the discussion of the principal theme of this study,[1] it is necessary to cast a brief glance over the origin and development of the meaning and use of the word milieu.

"Milieu" (*mi-lieu=medius locus*), originally signifying middle point or part, central place or portion, mid-point, center, had been employed in France as a term in physics at least as early as the seventeenth century (Pascal). The fourth edition of the dictionary of the French Academy[2] defines it as follows: "En termes de Physique, on appelle *Milieu*, Tout corps, soit solide, soit fluide, traversé par la lumière ou par un autre corps." In the fifth edition—1813—the following illustration in italics is added to the foregoing: "*La lumière se rompt différemment en traversant différens milieux.*"

"On appelle aussi *milieu*, Le fluide qui environne les corps. *L'air est le milieu dans lequel nous vivons. L'eau est le milieu qu'habitent les poissons.*"

Diderot's Encyclopedia[3] testifies to this same sense of "medium": "*Milieu*, dans la Philosophie mêchanique, signifie un espace matériel à travers lequel passe un corps dans son mouvement, ou en général, un espace matériel dans lequel un corps est placé, soit qu'il se meuve ou non.

"Ainsi on imagine l'éther comme un *milieu* dans lequel les corps célestes se meuvent.—L'air est un *milieu* dans lequel les corps se meuvent près de la surface de la terre.—L'eau est le *milieu* dans lequel les poissons vivent & se meuvent.—Le verre enfin est un *milieu*, en égard à la lumière, parce qu'il lui permet un passage à travers ses pores."

Auguste Comte[4] extended its signification as a term in biology to include "the totality of external conditions of any kind whatsoever":

"*Milieu* ..., non-seulement le fluide où l'organisme est plongé, mais, en général, *l'ensemble total des circonstances extérieurs d'un genre quelconque* the italics are ours, nécessaires à l'existence de chaque organisme déterminé. Ceux qui auront suffisamment médité sur le rôle capital que doit remplir, dans toute biologie positive, l'idée correspondante, ne me reprocheront pas, sans doute, l'introduction de cette expression nouvelle."

Hippolyte Taine who generalized it still further, broadened its connotation to comprehend the whole social surroundings.[5] Milieu as a *terminus technicus* is ordinarily considered as having been coined by Taine, but whether that be so or not, one may safely say that its wide acceptance is due, primarily, to him and to his renowned disciple Zola.[6]

In the course of the last century, the designation milieu became not only more generalized and more frequent in use, but also more extensive, and more specific and distinctive in meaning: "Depuis BALZAC who in 1841 in his *Comédie humaine, La maison du chat-qui-pelote*, préface, p. 2, used the term loosely, in the "vulgar" sense, le sens vulgaire du milieu social n'a fait que s'affirmer davantage par un emploi toujours plus généralisé: c'est devenu un cliché de la conversation de parler aujourd'hui d'un 'bon milieu,' d'un 'milieu intéressant,' etc."[7]

Littré[8] registers eighteen different definitions for the word milieu.

Friedrich Düsel[9] renders milieu by eighteen (18) German words.

In *Unsere Umgangssprache*,[10] milieu is translated into German by forty-six (46) words and phrases.

Claude Bernard, the celebrated French physiologist, differentiates between inner and outer milieu:[11] "Je crois ..., avoir le premier insisté sur cette idée qu'il y a pour l'animal réellement deux milieux: un milieu extérieur dans lequel est placé l'organisme et un milieu intérieur dans lequel vivent les éléments des tissus...." Probably as a result, we have today "micro-milieu" in micro-biology.

According to Jean Finot,[12] milieu "includes the sum total of the conditions which accompany the conception and earthly existence of a being, and which end only with its death."

The term milieu was introduced by Herbert Spencer into English literature as "environment," says Martin Schütze.[13] Although Carlyle employed the term "environment" as early as 1827,[14] nevertheless, the fact that the term is generally current, is undoubtedly attributable in the first place to Spencer.

The word "Umwelt" is quoted by J. H. Campe,[15] who believed himself to have been the coiner of the term; five years later (1816) Goethe used it at the beginning of his "Italienische Reise."[16]

The painstaking and scholarly German lexicographer, Daniel Sanders, who seldom fails to give his reader some reliable suggestion, refers in his *Wörterbuch der Deutschen Sprache*[17] (which despite the contributions of recent scholarship still remains a great work) to a passage in the poetical works of the Danish writer Baggesen (2, 102) in which the word "Umwelt" is employed. This passage occurs in the elegy entitled "Napoleon" addressed to Voß and written in 1800.[18] Baggesen, then, made use of "Umwelt" a decade before Campe.

Its Italian equivalent is "ambiente," which is noted here only because of the French "l'ambiance" and the English "ambient" and "circumambiency."

I
A Sketch of the History of the Idea of Milieu Down to the Nineteenth Century

Recorded mesologic[19] thinking begins with the ancient Jewish Prophets whose striking *aperçus* concerning the providential correspondence between the configuration of the surface of the earth and the destiny of nations, concerning the connection between "Landesnatur" and "Volkscharakter," etc., anticipated[20] a number of great thoughts of later anthropo-geographers.

Hippocrates (if he really is the author of the essay commonly ascribed to him and entitled περὶ αέρων ὑδάτων τόπων) investigates the effect of climate on man's nature, character, temperament, and life, with the emphasis on the regularity of the effect.[21] Owing to the imperfection of knowledge in his day, his observations are necessarily vague.[22] He limited himself to the problem of the relation between land and people.[23] He is said to be the founder of anthropo-geography.[24] His treatise is admirable and unequalled in the eyes of Auguste Comte.[25] Hippocrates, "in his work, *About Air, Water, and Places*, first discusses the influence of environment on man, physical, moral, and pathological. He divided mankind into groups, impressed with homogeneous characters by homogeneous surroundings, demonstrating that mountains, plains, damp, aridity, and so on, produced definite and varying types."[26]

Aristotle, in his *Politics*, enquires into the influence especially of geographical position on laws and the form of government,[27] while in his *Problems* he shows the far-reaching dependence of national character on the physical environment: "Zeigt ja doch Aristoteles selbst in einem andern Werke das entschiedenste Bestreben, eine sehr weitgehende Abhängigkeit des Volkscharakters von geographischen Verhältnissen zu erweisen. Während die Politik especially parts of the seventh book nicht über Andeutungen on the effect of the milieu hinausgeht discussed by Poehlmann, *l.c.*, on pp. 64–8, läßt der vierzehnte Abschnitt der 'Probleme,' welcher sich mit den Einwirkungen der Landesnatur auf Physik und Ethik des Menschen beschäftigt, deutlich einen Standpunkt erkennen, welcher auf das Lebhafteste an die physiologische Betrachtungsweise der neueren französisch-englischen Geschichtsphilosophie erinnert ..."[28]

Eratosthenes, in a work cited by Varro, sought to prove, in the opinion of the Italian scholar Matteuzzi prematurely, that man's character and the form of his government are subordinated to proximity or remoteness from the sun.[29] The greatest geographer of antiquity, Strabo, in his Geography, connected man with nature in a causal relation.[30]

John M. Robertson, noting that "theories of the influence of climate on character were common in antiquity," refers[31] to Vitruvius (VI, 1), Vegetius ("De re militari," 1, 2), and Servius (on Vergil, *Aeneid*, VI, 724). Ritter does not mention the effort of the ancients in this line of ideas.[32]

Giovanni Villani, the noted Florentine historian of the fourteenth century, observes with a deal of finesse that Arezzo by reason of its air and position produces men of great subtilty of mind.[33]

The Arabic statesman and philosopher of history, Ibn Khaldūn, little mentioned, yet known by his great work, the *Universal History*, attempted in the *Muqaddama*[34] (the preface, comprising the first volume of his *History*), which he composed between 1374 and 1378,[35] to explain the history and civilization of man, more especially of some of the Arabic peoples, by the encompassing physical and social conditions. The "First Section of the 'Prolegomena' treats of society in general, and of the varieties of the human race, and of the regions of the earth which they inhabit, as related thereto. It starts from the position that man is by nature a social being. His body and mind, wants and affections, for their exercise, satisfaction, and development, all imply and demand co-operation and communion with his fellows,—participation in a collective and common life....

"There follows a lengthened description of the physical basis and conditions of history and civilisation. The chief features of the inhabited portions of the earth, its regions, principal seas, great rivers, climates, &c., are made the subjects of exposition. The seven climatic zones, and the ten sections of each, are delineated, and their inhabitants specified. The three climatic zones of moderate temperature are described in detail, and the distinctive features of the social condition and civilisation of their inhabitants dwelt upon. The influence of the atmosphere, heat, &c., on the physical and even mental and moral peculiarities of peoples is maintained to be great. Not only the darkness of skin of the negroes, but their characteristics of disposition and of mode of life, are traced to the influence of climate. A careful attempt is also made to show how differences of fertility of soil—how dearth and abundance—modify the bodily constitution and affect the minds of men, and so operate on society....

"The Second Section of the 'Prolegomena' treats of the civilisation of nomadic and half-savage peoples.

"In it Ibn Khaldūn appears at his best, ... He begins by indicating how the different usages and institutions of peoples depend to a large extent on the ways in which they provide for their subsistence. He describes how peoples have at first contented themselves with simple necessities, and then gradually risen to refinement and luxury through a series of states or stages all of which are alike conformed to nature, in the sense of being adapted to its circumstances or environment."[36]

Ibn Khaldūn seems also to have had a clear idea of some aspects of the principle of relativity,[37] an integral part and inevitable concomitant of the theory of milieu, since "As causes of historians erring as they have done, there are mentioned by Khaldūn in the introduction the overlooking of the differences of times and epochs, …"[38]

About the middle of the sixteenth century we find Michelangelo avowing to Vasari (who hailed from Arezzo): "Any mental excellence

I may possess, I have because I was born in the fine air of your Aretine district."[39]

In "Measure for Measure" (Act III, Sc. I, v. 8–11), a play first produced in 1604, Shakespeare affirms of man:

> "... a breath thou art,
>
> Servile to all the skyey influences
>
> That do this habitation where thou keep'st,
>
> Hourly afflict."

During the Renaissance, Greek thought on milieu is resurrected in France. Thence it spreads later, particularly in the eighteenth century, to England and Germany. Jean Bodin bridges the gap existent since the close of classical antiquity. He is the first among modern writers not only to revive the idea in Western Europe,[40] but also to make it a subject for detailed investigation. Bodin thus first in French letters introduces and firmly establishes a line of study destined to be followed by a long list of authors among whom are to be found many illustrious French names.

Bodin "treats of physical causes with considerable fulness in the fifth chapter of the 'Method,'[41] and in a still more detailed and developed form in the first chapter of the fifth book of the 'Republic.'"[42] He traces the relation between climate and the ever changing fate of States, and elaborates the manifold effects of climate on States, laws, religion, language, and temperament.[43] In Bodin's view, man's physical constitution is closely and directly connected with climate and surrounding nature; it is in harmony with the behavior of the earth in the respective zones of his abode.[44] From this assumption of dependence of the human body on climate, there follow a number of inferences concerning the physical properties of man's constitution.[45] Temperament varies according to climate. Language, the generative power, diseases likewise depend indirectly on climate.[46] Man's talents and capacities do so no less.[47] The climate in each region always favors the development of some special aptitude;

on this basis he groups the peoples of the earth.[48] Although the nexus between human abilities and the physical milieu is thus intimate, yet reason, common to all men and invariable, is *per se* independent of physical environment.[49] He postulates, then, reason as the absolute part of the mind, not subject to surrounding influences, whereas the unfolding of the human faculties is relative to the environment. By taking this middle course concerning the effect of nature on man, Bodin escapes the extreme views of nature's compelling influence over man, on the one hand, and of man's total independence of nature, on the other.[50]

Bodin also investigates the influence upon national character of geographical situation, of elevation, of the quality of the native soil, and of an east-west position.[51] Nations and their civilizations differ according to the particular conditions of a given national existence.[52]

He holds fast to the doctrine of the freedom of the will. Man is morally free from environmental control. The circumambient medium determines only the *development* of man's capabilities.[53] Man can counteract, and may, even though with difficulty, overcome the injurious action of climate and nature.[54]

"... It is altogether unfair," concludes Flint,[55] "to put their general enunciations *i.e.*, those made by Hippocrates, Plato, Aristotle, Polybius, and Galen of the principle that physical circumstances originate and modify national characteristics, on a level with Bodin's serious, sustained, and elaborate attempt to apply it over a wide area and to a vast number of cases. Dividing nations into northern, middle, and southern,[56] he investigates with wonderful fulness of knowledge how climatic and geographical conditions have affected the bodily strength, the courage, the intelligence, the humanity, the chastity, and, in short, the mind, morals, and manners of their inhabitants; what influence mountains, winds, diversities of soil, &c., have exerted on individuals and societies; and he elicits a vast number of general views...."

Bodin, "der größte theoretische Politiker Frankreichs im 16. Jahrhundert," declares Renz,[57] "besitzt ... das unbestreitbare

Verdienst, wenn nicht die Grundgedanken und nicht ausschließlich originale Gedanken, so doch die erste weitgehende wissenschaftliche Untersuchung über den Zusammenhang zwischen umgebender Natur und Menschenwelt in neuerer Zeit auf dem Boden der Erfahrung und Wissenschaft des 16. Jahrhunderts angestellt zu haben."

Bodin, "writing in 1577 OF THE LAWES AND CUSTOMES OF A COMMON WEALTH (English edition translated by Richard Knowlles 1605), contains, as Professor J. L. Myres has pointed out (Rept. Brit. Assoc., 1909 1910, p. 593), 'the whole pith and kernel of modern anthropo-geography....'"[58] And Renz believes that "In der Bodinschen Behandlung der Theorie des Klimas finden sich die Anfänge der Anthropogeographie und der Ethnographie..."[59]

Writing in 1713, Lenglet du Fresnoy, toward the end of the sixth chapter of the first volume of his *Méthode pour étudier l'histoire*, expresses, decades before Montesquieu, the latter's basic idea of the effect of social and political milieu on laws.[60]

In any discussion of milieu, Montesquieu is the writer most frequently mentioned, although not the most often read and quoted. He devotes the well-known five "Books," from the fourteenth to the eighteenth, of his magnum opus, *L'Esprit des Lois* (1748),[61] to a consideration of this idea which, as has already been seen, was anything but original with him.[62] In Books fourteen to seventeen he treats of the relation of laws to climate, and in Book eighteen of their relation to soil. In the fourteenth[63] he discusses the effect of climate on the body (and mind) of individual man, in the fifteenth[64] on civil slavery, in the sixteenth[65] on domestic slavery, in the seventeenth[66] on political servitude, and lastly in the eighteenth[67] he delineates the influence of the fertility and barrenness of the soil. By climate he means little more than heat and cold. In the light of the continued high praise bestowed on him for much longer than a century, the altogether too general and dogmatic statements of these short seventy-odd pages would seem somewhat meager, so that upon their perusal one is very likely to suffer an outright disenchantment. Therefore, Flint's judgment appears overdrawn, when he says that

Montesquieu "showed on a grand scale and in the most effective way ... that, like all things properly historical, they laws, customs, institutions must be estimated not according to an abstract or absolute standard, but as concrete realities related to given times and places, to their determining causes and condition, and to the whole social organism to which they belong, and the whole social medium in which they subsist. Plato and Aristotle, Machiavelli and Bodin, had already, indeed, inculcated this historical and political relativism; but it was Montesquieu who gained educated Europe over to the acceptance of it."[68]

Turgot's sketch of a 'Political Geography' shows "that he had attained to a broader view of the relationship of human development to the features of the earth and to physical agencies in general than even Montesquieu. And he saw with perfect clearness not only that many of Montesquieu's inductions were premature and inadequate, but that there was a defect in the method by which he arrived at them.... The excellent criticism of Comte, in the fifth volume of the 'Philosophie Positive,' and in the fourth volume of the 'Politique Positive,' on this portion of Montesquieu's speculations, is only a more elaborate reproduction of that of Turgot, and is expressed in terms which show that it was directly suggested by that of Turgot."[69]

Cuvier "had not hesitated to trace the close relation borne by philosophy and art to the underlying geological formations."[70]

In the teaching of a number of great thinkers of the seventeenth and eighteenth centuries, man is "the product of environment and education" and, in their opinion, "all men were born equal and later became unequal through unequal opportunities."[71]

Goethe echoed Herder's thought when he remarked to Eckermann on the flora of a country and the disposition of its residents: "Sie haben nicht Unrecht, sagte Goethe (d. 2. April 1829), und daher kommt es denn auch, daß man der Pflanzenwelt eines Landes einen Einfluß auf die Gemütsart seiner Bewohner zugestanden hat. Und gewiß! wer sein Leben lang von hohen ernsten Eichen umgeben wäre, müßte ein anderer Mensch werden, als wer täglich unter luftigen Birken sich erginge..."[72] And again, when he said of environment and national character: "... so viel ist gewiß, daß außer dem Angeborenen der Rasse, sowohl Boden und Klima als Nahrung und Beschäftigung einwirkt, um den Charakter eines Volkes zu vollenden ..."[73] And in the following, Goethe but reiterates Herder's oft uttered admiration for islanders and coast dwellers: "Auch von den Kräften des *Meeres* und der *Seeluft* war die Rede gewesen (d. 12. März 1828), wo denn Goethe die Meinung äußerte, daß er alle Insulaner und Meer-Anwohner des gemäßigten Klimas bei weitem für produktiver und tatkräftiger halte als die Völker im Innern großer Kontinente."[74] And: "Es ist ein eigenes Ding, erwiederte Goethe (d. 12. März 1828),—liegt es in der Abstammung, liegt es im Boden, liegt es in der freien Verfassung, liegt es in der gesunden Erziehung,—genug! die Engländer überhaupt scheinen vor vielen anderen etwas voraus zu haben ..."[75]

Wolf and Niebuhr began to examine historical *sources* "nach neuen Prinzipien des Eingetauchtseins in eine bestimmte seelische Umwelt, in ein klargezeichnetes zeitgenössisches Milieu."[76]

One of the principal offices of an historian, according to August Wilhelm Schlegel, is "Die zeit- und kulturgeschichtliche Bedingtheit aller Erscheinungen aufzuzeigen."[77] But the effect of physical milieu on history is not rated high in the philosophy of the romanticists.[78]

Ingeniously, albeit not with his wonted acuteness, Hegel penned the concept "Volksgeist."[79] The saying, which now seems trivial, that every nation and every man in the nation is "ein Kind seiner Zeit," is said to be Hegel's.[80] Hegel, however, distinctly rejected the idea of explaining "die Geschichte und den Geist der verschiedenen Völker aus dem Klima ihrer Länder."[81] The implication would be that one single factor might satisfactorily be held responsible for all progress in human history. As climate can not explain everything to Hegel, it seems not to explain anything at all to him. Hegel, then, is excessive in his denial of the power of environment. This is markedly shown by his thinking his position substantiated by the fact that the climate of Greece, although the same since classical antiquity, has not changed the Turks who now *i.e.*, early in the nineteenth century dwell in Greece into ancient Greeks.[82]

II
A Sketch of the History of the Idea of Milieu Since the Beginning of the Nineteenth Century

Anthropo-geography, Geography and History

The theory of social environment, as we have seen, gradually rises, especially since the renaissance, parallel with the theory of physical milieu. The stream of thought commences to broaden on both sides as we approach the eighteenth century, and broadens still further, and deepens, in the nineteenth, when specialization occurs or continues in anthropo-geography, biology, jurisprudence and economics, anthropology, sociology, and literature, and latterly in physics. These furnish us the divisions for subsequent discussions.[83]

All antecedent thought on the subject converges in Herder and from this focal point, as a collecting and fructifying center, it emerges, branches out and radiates in a definite number of directions. This can only be indicated here.[84] One main ramification leads us to anthropo-geography. Consequently, we must now turn to a detailed consideration of the idea of milieu in anthropo-geography.[85]

Karl Ritter first in anthropo-geography elucidated Herder's ideas on environment. "... KARL RITTER steht auf HERDERS Schultern, wenn er in seiner 'Allgemeinen Erdkunde' den Gedanken der tiefgehenden Beeinflussung der Völkergeschichte durch die äußeren Umgebungen entwickelt ..."[86] Ritter is said to be given too much credit for connecting scientifically geography and history: "C. Ritter führte, ... die Herder'schen Anschauungen deutlicher aus. Die wissenschaftliche, nicht bloß äußerliche Verbindung von Geographie und Geschichte kettet sich an seinen Namen. Nicht ganz mit Recht; ..."[87] Richthofen thinks that Ritter's basic idea was almost without influence on geography; only the historians profited by it.[88]

Alexander von Humboldt, on the other hand, declares in the first volume of his *Cosmos* that "The views of comparative geography have been specially enlarged by that admirable work, Erdkunde im Verhältnis zur Natur und zur Geschichte, in which Carl Ritter so ably delineates the physiognomy of our globe and shows the influence of its external configuration on the physical phenomena on its surface, on the migrations, laws, and manners of nations, and on all the principal historical events enacted upon the face of the earth."[89]

In the *Erdkunde*,[90] Ritter propounds a program for anthropo-geographical investigation, i.e., for the investigation of the mutual relation between man and his environment. As every moral man should, so should also "jeder menschliche Verein, jedes Volk seiner eigenen inneren und äußeren Kräfte, wie derjenigen der Nachbarn und seiner Stellung zu allen von außen herein wirkenden Verhältnissen inne werden."[91] Nature exercises greater influence over peoples than over individual men: "Die Eigentümlichkeit des Volkes kann nur aus seinem Wesen erkannt werden, aus seinem Verhältnis zu sich selbst, zu seinen Gliedern, zu seinen Umgebungen, und weil kein Volk ohne Staat und Vaterland gedacht werden kann, aus seinem Verhältnis zu beiden und aus dem Verhältnis von beiden zu Nachbarländern und Nachbarstaaten. Hier zeigt sich der Einfluß, den die Natur auf die Völker, und zwar in einem noch weit höheren Grade, als auf den einzelnen Menschen ausüben muß ...

"Denn durch eine höhere Ordnung bestimmt, treten die Völker wie die Menschen zugleich unter dem Einfluß einer Tätigkeit der Natur und der Vernunft hervor aus dem geistigen wie aus dem physischen Elemente in den Alles verschlingenden Kreis des Weltlebens. Gestaltet sich doch jeder Organismus dem inneren Zusammenhange und dem äußeren Umfange nach ... Sie (Völker und Staaten) stehen alle unter demselben Einflusse der Natur ..."[92] To the problem of the reciprocal relation between external and internal factors, Ritter devoted a special essay, entitled "Über das historische Element in der geographischen Wissenschaft," which he read before the Academy of Sciences at Berlin in 1833.[93]

In Alexander von Humboldt's *Ansichten der Natur*,[94] "Everywhere the reader's attention is directed to the perpetual influence which physical nature exercises on the moral condition and on the destiny of man."[95] In passing, Humboldt also touches on environment in the first volume of his chef-d'oeuvre, *Kosmos*, assigning it, however, but a modest rôle: "Es würde das allgemeine Naturbild, das ich zu entwerfen strebe, unvollständig bleiben, wenn ich hier nicht auch den Mut hätte, das Menschengeschlecht in seinen physischen Abstufungen, in der geographischen Verbreitung seiner gleichzeitig vorhandenen Typen, in dem Einfluß, welchen es von den Kräften der Erde empfangen und wechselseitig, wenn auch schwächer, auf sie ausgeübt hat, mit wenigen Zügen zu schildern. Abhängig, wenn gleich in minderem Grade als Pflanzen und Tiere, von dem Boden und den meteorologischen Prozessen des Luftkreises, den Naturgewalten durch Geistestätigkeit und stufenweise erhöhte Intelligenz, wie durch eine wunderbare sich allen Klimaten aneignende Biegsamkeit des Organismus leichter entgehend, nimmt das Geschlecht wesentlich Teil an dem ganzen Erdenleben."[96]

J. G. Kohl's book, *Der Verkehr und die Ansiedlungen der Menschheit in ihrer Abhängigkeit von der Gestaltung der Erdoberfläche*,[97] occupies itself with the question of the dependence of human progress in general, and of density and concentration of population in particular, upon natural conditions. The causes of these phenomena are, to Kohl, partly moral or political, and partly physical. The physical causes of concentration are twofold: "Teils sind es solche, die von dem mehr oder minder großen Produktenreichtum des Bodens, teils solche, die von der Gestaltung der Erdoberfläche abhängen ... so zeigt sich dann, daß von allen verschiedenen Ursachen der Kondensierung der Bevölkerung die Bodengestaltung die allerwichtigste ist."[98] Opposed to these natural conditions is a series of what Kohl styles political influences, such as national character, institutions created by the State, laws, etc.—"Die moralischen oder politischen Ursachen der verschiedenen Dichtigkeit der Bevölkerung sind in dem Kulturzustande und besonders in der politischen Verfassung der Bewohner der verschiedenen Erdstriche begründet ... Auch sind viele verschiedene Sitten der Völker als einflußreiche Ursachen der mehr oder minder großen Dichtigkeit der Bevölkerung zu betrachten."[99] Not only national character, but also education is to be counted among the political influences: "Unter politischen und moralischen Einflüssen, die nicht von der Natur bedingt werden, verstehen wir solche Kräfte, solche Volkstalente und Eigentümlichkeiten des Charakters, die nicht der Boden, die Luft und das Klima dem Volke geben. So groß nämlich auch die Gewalt des Bodens, des Klimas und der Natur ist, so sehr die Zonen, die Gebirge, die Sümpfe, die Wälder, die Wüsten u.s.w. alle Bevölkerung, die in ihre Gebiete fällt, auf einerlei Weise zu bilden und zu modeln streben, so sehr behauptet doch immer noch nebenher der ursprüngliche Charakter des Stammes und die Erziehung, welche das Volk sich gibt, ihre eigenen Rechte. Es existieren beide Einflüsse neben einander, beschränken sich gegenseitig, aber sie heben sich nicht auf ... Das, was nun nicht vom Boden abhängt und was ein Volk auf jeden Boden, den es bezieht, mit hin bringt, ist wiederum Zweierlei, entweder etwas Angeborenes oder etwas Angenommenes."[100] It is difficult to differentiate between what is due to original endowment and what to the milieu, yet natural influences can not be ignored: "Welcher Geist ... möchte den Versuch wagen, zu entscheiden, was im Charakter des Volkes ... Angenommenes und was Selbstgegebenes sei, was endlich in ihren Handlungen und Bewegungen von Klima und Landesbeschaffenheit bedingt werde. Die Charaktergepräge der Nationen, wie wir sie jetzt in diesen neuesten Momenten der weltgeschichtlichen Entwicklung sehen, sind Gebilde, welche unter der Einwirkung unerforschbar vielfacher Einflüsse entstanden sind.... Und doch stehen sie (die Natureinflüsse, die von den Historikern gewöhnlich unberücksichtigt geblieben sind) vielleicht auch bei allen jenen Dingen, die wir im Vordergrunde agieren sehen, im Hintergrunde und wirken als die Quellen der Erscheinungen mittelbar selbst da, wo wir dieselben anderen Ursachen zuschreiben. So mag jede Art der Staatsverfassung, der Gewerbzweige geschöpft und hervorgeblüht sein aus der Tiefe des

Nationalgeistes, des Boden- und des Luftgeistes, während wir sie als Willkürliches und Selbstgegebenes auffassen."[101]

The naturalist Karl Ernst von Baer discusses the influence of external nature upon the social relations of individual nations and upon the history of mankind in general,[102] while the geologist Bernhard Cotta attempts to show the effect of soil and geological structure on German life.[103] Accepting, in the main, Cotta as a basis, J. Kutzen, in *Das deutsche Land, Seine Natur in ihren charakteristischen Zügen und sein Einfluß auf Geschichte und Leben der Menschen, Skizzen und Bilder*,[104] the bulk of which book is physical geography, intersperses therewith anthropo-geographical statements that are in some cases interwoven in, and in others added to, the descriptive parts, pointing out the relation of environment to the life and history of the Germans.[105] Kutzen claims his work to be the first that treats the *whole* of Germany in the way just indicated.

In The Natural History of the German People,[106] W. H. Riehl studies the action of natural conditions on man. He is concerned with the connections between land and people: "Will man die naturgeschichtliche Methode der Wissenschaft vom Volke in ihrer ganzen Breite und Tiefe nachweisen, dann muß man auch in das Wesen dieser örtlichen Besonderungen des Volkstumes eindringen. In der Lehre von der bürgerlichen Gesellschaft ist das Verhältnis der großen natürlichen Volksgruppen zueinander nachgewiesen: hier sollen diese Gruppen nach den örtlichen Bedingungen des Landes, in welchem das Volksleben wurzelt, dargestellt werden. Erst aus den individuellen Bezügen von LAND UND LEUTEN entwickelt sich die kulturgeschichtliche Abstraktion der bürgerlichen Gesellschaft."[107] And "Das vorliegende Buch hat sich das bescheidenere Ziel gesteckt, zusammenhängende Skizzen zu liefern zur Naturgeschichte des Volkes *in seinem Zusammenhang mit dem Lande*."[108] His chief aim is to prove that the connection between land and people is the basis of all social development and of all social research: "Ich hatte mir von Anbeginn das Ziel gesteckt, den Zusammenhang von Land und Volk als Fundament aller sozialen und politischen Entwicklung, als

Ausgangspunkt aller sozialen Forschung nachzuweisen, und dieses Hauptziel, die eigentliche Tendenz des Buches, hat heute noch denselben Wert, dieselbe fördernde Kraft wie vor einem Menschenalter."[109] He wants to show how "Volksart" and "Landesart" hang together, how nationality grows organically out of the soil: "Ich nenne dieses Wanderbuch einen zweiten Band zu 'Land und Leuten.' In jener Schrift verarbeite ich zahlreiche Wanderskizzen, um den Zusammenhang von Volksart und Landesart, das organische Erwachsen des Volkstumes aus dem Boden nachzuweisen."[110] Everywhere Riehl finds "an organic relation between nature and man," according to Gooch.[111] Riehl recognizes "that man could only develop within the limits imposed by nature."[112] The problem of how locality affects social groups has, of course, not originated with Riehl, but it received a reformulation at his hands. It must be added, however, that his bombastic assertions far outrun his data. His claims are disproportionate to his facts.[113]

Alfred Kirchhoff brilliantly sketches the reciprocal relations between land and people in Germany, in an essay entitled *Die deutschen Landschaften und Stämme*.[114]

Achelis[115] refers to Bastian's doctrine of geographical provinces, "wo eine Reihe rein physikalischer Agentien: Temperatur, Boden, Flora, Fauna, etc. sich mit entsprechenden psychischen kombinieren, so daß man in konzentrischer Reihenfolge von botanischen, zoologischen und anthropologischen Kreisen reden könnte. Der leitende Grundsatz, sagt Bastian, für geographisch-typische Provinzen fällt in die Abhängigkeit des Organismus von seiner geographischen Umgebung (*le Milieu* oder *Monde ambiant*), in eine gegenseitig festgeschlossene Wechselwirkung und also in Naturgesetze, mit denen sich rechnen läßt (*Zur Lehre von den geographischen Provinzen* Berlin, 1886, S. 6)."

The reciprocal influences of man and his environment are illustrated by Alfred Kirchhoff in *Mensch und Erde, Skizzen von den Wechselbeziehungen zwischen beiden*.[116]

Ferdinand von Richthofen[117] traces the gradual evolution of "Siedlung und Verkehr," under which two concepts he subsumes all relations of man to the soil.[118]

It was Friedrich Ratzel, however, who "performed the great service of placing anthropo-geography on a secure scientific basis. He had his forerunners in Montesquieu,[119] Alexander von Humboldt, Buckle, Ritter, Kohl, Peschel and others; but he first investigated the subject from the modern scientific point of view, … and based his conclusions on world-wide inductions, for which his predecessors did not command the data."[120] He "has written the standard work on *Anthropogeographie*."[121] Employing the analytical method, Ratzel was the first to divide the subject-matter into categories: "Ratzel hat das Verdienst, daß er zuerst den Stoff in Kategorien teilte. Er wendet die analytische Methode der allgemeinen Geographie an und betrachtet den Einfluß einzelner Naturgegebenheiten auf den Menschen, z.B. der Inseln, Halbinseln, Gebirge, Ebenen, Steppen, Wüsten, Küsten, Flußmündungen[122] usw. Die analytische Methode allein kann zum Ziele führen."[123] The great and permanent merit of Ratzel's *Politische Geographie*[124] is its setting forth how closely the State is bound to the physical milieu.[125] It treats partly of the effect of nature and soil on the formation of the State and on political boundaries.[126] Ratzel expounds environmental action also in his books *Die Vereinigten Staaten von Amerika*,[127] *The History of Mankind*,[128] and in his article on "The Principles of Anthropo-geography."[129] Among his followers is to be counted Andrew R. Cowan, whose *Master-Clues in World-History*[130] is "deeply impregnated with Ratzel's teachings."[131] Camille Vallaux devotes the fifth chapter (pp. 145–73) of his *Géographie Sociale, Le Sol et L'État*,[132] to a criticism of the theories of *Raum* (space) and of Lage (situation) as developed by Ratzel in his *Politische Geographie*. And, in general, Ratzel's "published work had been open to the just criticism of inadequate citation of authorities."[133] O. Schlüter in "Die leitenden Gesichtspunkte der Anthropogeographie, insbesondere der Lehre Friedrich Ratzels"[134] gives us the best single estimate of Ratzel, the best orientation—within the compass of an article well written, well poised, and illuminating—on Ratzel's work, thought, method, and application.[135]

Geography and History

We shall now see, first, the stand taken by some French writers, and then that taken by German and English writers, on the question of how physical environment affects history.

One of the "three most philosophical writers on climate,"[136] Charles Comte, not related by birth to the founder of Positivism, is, likewise, one of the earliest disciples of Herder in France. Herder "seems to have helped to inspire"[137] Charles Comte's *Traité de Législation*.[138] Charles Comte's "discussion of the questions which relate to the influence of physical nature on human development must have been the fruit of long and careful study. It was as great an advance on Montesquieu's treatment of the subject as Montesquieu's had been on that of Bodin. It disproved, corrected, or confirmed a host of Montesquieu's observations and conclusions. It showed that he had ascribed too much to climate, and too little to the configuration of the earth's surface, the distribution of mountains and rivers, &c.; and that he had conceived vaguely, and even to a large extent erroneously, of the modes in which climate and the fertility or sterility of soil affect human development. But while Comte thus justly criticised Montesquieu, he himself exaggerated the efficiency of physical agencies. Indeed, he virtually traced to their operation the whole development of history ... he has assumed that physical agencies ultimately account for historical change and movement, for public institutions and laws....

"Charles Comte fully recognises that the same physical medium has a very different influence on different generations; and that institutions and laws, education and manners, and, in a word, all the constituents of the social medium, have as real an influence on the development of history as those of the physical medium. Yet he assumes the latter to be the first, although to a large extent only indirect, causes of the whole amount of change effected."[139]

Victor Cousin, another Frenchman, reconnects with Herder. Cousin had direct acquaintance with at least the principal work of Herder, for the rendering of whose "Ideen" into French by Quinet he seems responsible.[140] In the eighth lecture of his "admired"[141] *Cours de 1828 sur la Philosophie de l'Histoire*, he discourses on the rôle that geography plays in history.

F. Guizot, in the fifth lecture of *The History of Civilization*,[142] comments briefly on the influence of external circumstances upon liberty.

The romantic French historiographer, Jules Michelet, in his *Histoire de France* (second volume, 1833), and in his *Histoire Romaine* (1839), interlinks geography with history, and brilliantly describes the countries whose histories he is writing. Like some before him (such as Montesquieu), and many after him (such as Riehl, Curtius, and Gothein),[143] who traveled in the respective countries before describing them or composing their history, Michelet, as one preliminary measure toward equipping himself for such a task, visited Italy[144] and various parts of France, the latter repeatedly, in order to gain a first hand impression of the physical milieu and the people of those lands. He is said to be the first *sic!* in France who, under the influence of Herder, had the idea that geography was the foundation of history: "Sous l'influence de Herder, il Michelet eut, le premier en France, l'idée que la géographie était le fondement de l'histoire: 'Le matériel, la race, le peuple qui la continue me paraissaient avoir besoin qu'on mît dessous une bonne et forte base, la terre, qui les portât et qui les nourrît. Et notez que ce sol n'est pas seulement le théâtre de l'action. Par la nourriture, le climat, etc., il y influe de cent manières. Tel le nid, tel l'oiseau. Telle la patrie, tel l'homme.'"[145] Without this basis, the actor in history, the people, would be treading on air like figures in some Chinese paintings. Says Jules Simon of the celebrated tableau in the second volume of the *Histoire de France*: "Son héros Michelet's ... c'est la France. Il en fait une description qui remplit tout le troisième livre et qui est un chef-d'oeuvre. Chose nouvelle, cette géographie a autant de mouvement que l'histoire. Elle est animée, vivante, agissante. Il en montre à merveille l'utilité, la nécessité. Sans cette base géographique, le peuple, l'acteur historique, semblerait marcher en l'air, comme dans les peintures chinoises, où le sol manque."[146] In the *Introduction to Universal History* (1831), Michelet says, "In Germany and Italy, fatality is still strong; moral freedom is still borne down by powerful influences of race, locality, and climate."[147]

Ernst Kapp, in the *Philosophische Erdkunde*,[148] criticizes writers on the philosophy of history for their failure to give due attention to the geographical existence of the nations. Nor are geographical intermezzos alone sufficient: "Man these writers hat zwar eine Ahnung von dem geographischen Element in der Geschichte, nicht aber das deutliche Bewußtsein, daß die Menschheit an dem Planeten ihre physische Individualität besitzt, daß sie zu ihm sich verhält, wie die Seele zum Leib. Anstatt die geographische Betrachtung durch und durch mit der historischen verwachsen zu lassen which he proposes to do, hat man teils geographische Intermezzos nach subjektivem Gutdünken ... eingestreut, teils auch sich mit einer dem Ganzen voraufgeschickten geographischen Grundlage ein für allemal begnügt. Man hat hierbei nicht bedacht, daß man die Geschichte, wenn man ihr den planetarischen Grund und Boden, auf den man sie von vornherein stellt, wegrückt, zwischen Himmel und Erde schweben läßt und ihre Behandlung dem veränderlichen Luftzuge des subjektiven Beliebens mehr oder minder preisgibt ... Darin ruht die Selbständigkeit der geographischen Wissenschaft, ..., daß ihr Objekt die Erde ist, ... die Erde, wie sie bestimmend auf die Entwicklung des Geistes einwirkt und hinwiederum vom Geist bestimmt und verändert wird. Dies Verhältnis des Planeten zum Geist ist ein wesentliches."[149]

Arnold H. Guyot, "ce Suisse transplanté en Amérique,"[150] treats the same topic in the *Géographie physique comparée, considérée dans ses rapports avec l'histoire de l'humanité.*[151]

The frequently misquoted Henry Thomas Buckle, in the celebrated second chapter of the *History of Civilization in England*,[152] shows the largely indirect effects of climate, food, and soil, chiefly upon the civilizations—of India, Egypt, Mexico, Peru, etc.—anterior to those of Europe, and of a fourth class of physical agents, namely, of what he terms the general aspect of nature upon the imagination—religion, literature, art—of those peoples. Buckle does not maintain that these four classes of the Environment were the *sole* factors in producing civilization; in fact he makes it quite clear that they were *not* the only factors, that they affected the civilizations mentioned in an indirect way and he indicates how this has taken place. Buckle's statements of his ideas had been misrepresented, twisted, and distorted to such a degree that John M. Robertson felt impelled to write a whole book[153] in rebuttal, in order to set Buckle's detractors and controversial critics right and to refute their unfair imputations to Buckle's intended meaning.

The romanticist Ernst Curtius is sometimes referred to as one of those historians who give adequate expression to the action of the physical milieu upon the course of history. But Vallaux declares that Curtius, like Michelet, has made of human geography and of political geography *merely* a preliminary and introductory science to history: "une science auxiliaire ou plutôt liminaire, sorte de *portique d'entrée* the italics are ours pour leurs brillantes constructions,"[154] lending thus support to Kapp's contention.[155] Nor would Ratzel be content with a portrayal of the land as an introduction to the history of a country, even though it be as richly colored as that drawn by Curtius.[156] A description, in itself, fails to penetrate to the core of the relation. If we now turn to Curtius' *The History of Greece*,[157] we find that the first chapter in the first book[158] considers Land and People, a part of which (pp. 9–18) gives a geographical description of Hellas, and another part of which (pp. 19–25, seven pages scant) points out the connection between the land and the people. Elsewhere,[159] Curtius shows the interaction between the physical environment of Athens and the Athenians.[160]

George Grote, whose account of the relation between the Greek land and the Greek people is held by some[161] to be excellent, in *A History of Greece*,[162] devotes four pages (227–30) of the chapter on General

Geography and Limits of Greece to show the effects of the configuration of Greece upon the political relation of the inhabitants[163] and the effects upon their intellectual development,[164] the rest of the chapter being given over to a description of the geography of Greece.

Alfred E. Zimmern, in *The Greek Commonwealth, Politics and Economics in Fifth-Century Athens*,[165] deals very cleverly with the main features of the material environment of Greek civilization: The Mediterranean Area; The Sea; The Climate; The Soil; Fellowship, or the Rule of Public Opinion, under which headings he discusses the influence of environment upon Greek institutions.[166]

As early as 1864, G. P. Marsh investigates the subject of man's reaction on his milieu in *Man and Nature, or Physical Geography as Modified by Human Action* (London).

John William Draper, in his *History of the Intellectual Development of Europe*,[167] in the composition of which Herderian ideas were the guides,[168] first attempts to show (vol. I, pp. 6–17) that individual man, as well as communities, nations, and universal humanity, are under the control of physical conditions; then (pp. 23–35) he points out how the topography, meteorology, and secular geological movements of Europe affected its inhabitants. On the whole, he overstates the force of environment and neglects the human factor; nevertheless his uncompromising affirmations bring out strikingly some of the environmental effects on man.

The uncritical Max Duncker, in the nine volume *Geschichte des Altertums*,[169] not only has chapters on *Land und Volk*, or *Land und Stämme* at the beginning of the history of a given nation, but he also dwells elsewhere in his text on the sway of geography in history.

Élisée Réclus, in the magistral *Nouvelle Géographie Universelle* (1879 ff.), speaking of the difficulties encountered by research, queries: "... Was verdanken die Nationen dem Einfluß der Natur, die sie umgibt? Was verdanken sie dem Milieu, das ihre Vorfahren bewohnten, ihren Rasseinstinkten, ihren verschiedenartigen Mischungen, den von Außen eingeführten Überlieferungen? Man weiß es nicht, kaum daß einige Lichtstrahlen in jene Finsternis dringen."[170] The preponderance of European nations is by no means attributable, as some arrogantly and self-conceitedly fancied, to any racial endowment; on the contrary, it is due to the favoring conditions of the physical environment prevailing in Europe: "Man weiß, wie mächtig der Einfluß des geographischen Milieu auf die Fortschritte der europäischen Nationen gewesen ist. Ihre Überlegenheit ist keineswegs, wie einige sich dünkelhafter Weise eingebildet haben, der eigentümlichen Anlage der Rassen zuzuschreiben, denn in anderen Gegenden der alten Welt haben sich eben dieselben Rassen weniger schöpferisch erwiesen. Es sind die glücklichen Bedingungen der Wärme, des Klimas, der Gestalt und Lage des Festlandes, welche den Europäern die Ehre verschafft haben, die ersten gewesen zu sein in der Kenntnis der Erde in ihrem ganzen Umfange und lange Zeit an der Spitze der Zivilisation geblieben zu sein."[171] These conditions help to explain, in part, the character of the nations: "Mit vollem Recht lieben es also die historischen Geographen bei der Gestalt der verschiedenen Erdteile und bei den Folgen zu verweilen, welche sich daraus für die Bestimmung der Völker ergeben. Die Gestalt der Hochebenen, die Höhe der Berge, der Lauf und der Reichtum der Flüsse, die Nachbarschaft des Ozeans, die Gliederung der Küsten, die Temperatur der Atmosphäre, die Häufigkeit oder Seltenheit des Regens, die unzähligen wechselseitigen Einflüsse der Sonne, der Luft und der Gewässer, alle Erscheinungen des Pflanzenlebens habe eine Bedeutung in ihren Augen und dienen ihnen (wenigstens zum Teil), den Charakter und das erste Leben der Nationen zu erklären ..."[172] Continental and oceanic forms and other features of the globe vary in their value for man in accordance with the stage of civilization to which he attained.[173] Notwithstanding this separation, in principle, of natural and national influences upon social evolution, its application to concrete cases Réclus finds arduous: "Durch das Studium der Sonne und durch die unablässige Beobachtung der klimatischen Erscheinungen können wir ganz allgemein verstehen, welches der Einfluß der Natur auf die Entwicklung der Völker gewesen ist; aber es ist schwieriger, das auf jede Rasse, auf jede Nation zu verteilen...."[174]

P. Mougeoulle's theory in *Les problèmes de l'histoire*,[175] is an altogether one-sided geographical theory of history.[176] The sole cause of the external as well as the internal history of peoples, is, in his opinion, the geographical Milieu.[177] To Mougeoulle, the Milieu is the author, whereas man is the actor of the Drama of history.[178]

Léon Metchnikoff, in *La Civilisation et Les Grands Fleuves Historiques*,[179] pays some attention to the influences (astronomic, physical—the geosphere, the hydrosphere, and the atmosphere—, vegetal, animal, anthropological) of the milieu on man and society; yet his main care is with the action of parts of the hydrosphere on human progress. Following C. Böttiger (*Das Mittelmeer*, Leipzig, 1859), Metchnikoff distinguishes the three milieus: fluvial or potamic, mediterranean or thalassic, and oceanic or universal.[180] On this basis he divides universal history into three periods: 1) the period of the fluvial civilizations (temps anciens), furnishing the principal theme of his argument (discussed in the last four chapters of his book); 2) that of the mediterranean civilizations (temps moyens); 3) and that of the oceanic civilizations. The fluvial or ancient period, from the beginnings to *circa* 800 B.C., comprises the history of the four great civilizations of antiquity, in Egypt, Mesopotamia, India, China, "qui ont eu pour milieu géographique des régions arrosées par certains fleuves ou couples de fleuves célèbres." The mediterranean or middle period extends from the seventh century B.C.—the foundation of Carthage—to Charles the Fifth. The modern or oceanic period has two epochs: a) the *atlantic* epoch, from the discovery of America to about the middle of the nineteenth century; and b) the *universal* epoch, just beginning.[181] In the main, Metchnikoff limits the scope of his work to the compass of fluvial civilizations. He studies in detail the four great historical rivers or pairs of rivers (the Nile, the Tigris and the Euphrates, the Indus and the Ganges, and the Hoangho and the Yangtze-Kiang, those great educators of mankind) in their bearing upon the four grand civilizations—Chinese, Hindu, Assyro-Babylonian, and

Egyptian—of remote antiquity, all of which expanded in fluvial regions.[182] The River, in all countries, presents itself to Metchnikoff as the living synthesis of all the complex conditions of the climate, of the soil, of the configuration of the earth, and of the geologic formation. In Egypt and in China, in India and in Mesopotamia, the River has been "comme une synthèse vivante des conditions géographiques les plus multiples."[183] He finds that each of the four great monarchies of antiquity had been a natural consequence or result of the hydrological system of the country that served as its cradle, and that history, in the entire ancient world, had been a toil, a forced labor ("une corvée"), imposed on a part of mankind by certain orographic peculiarities of the Milieu. Metchnikoff concludes that in these empires "le Milieu s'est trouvé être invariablement le vrai créateur de l'histoire." The eloquent example of these four grand ancient civilizations sufficiently proves to him that no important historical expansion could ever occur in any country of the world, unless the milieu condemned its inhabitants to that excessive solidarity which he shows to have been brutally imposed everywhere at the shores of these great historical rivers; a milieu is conceivable, however, where this condition, rigorously required by history, may be fulfilled by an environmental factor other than a river or a system of rivers.[184] Metchnikoff protests that he is far from advocating potamic[185] or geographical[186] fatalism.[187]

Babington's study of the power of environment over history points out the fallacy of the race theory in the history of the Roman empire, of Germany, and of China.[188]

N. S. Shaler, in *Nature and Man in America*,[189] traces, on the one hand, the action of environment on organic life, and, on the other, the effect of geographic conditions on the development of peoples, more especially on that of man in North America.[190]

Since about the middle of the eighties, under the leadership of the late historian E. A. Freeman and of the illustrious statesman and scholar, Lord James Bryce, "a marked revival of interest" has been exhibited in England in studying the physical milieu as it relates to man and human society, institutions and history.[191]

The leading point of view in H. F. Helmolt's *The History of the World, a Survey of Man's Record*,[192] is the treatment of man's relation to his physical environment, the relation of geography to history, the dependence of man on his geographical surroundings. "It Helmolt's *History* deals with history in the light of physical environment.... Its ground plan, so to speak, is primarily geographical...."[193] It was conceived in the spirit of Ratzel;[194] it is said to have brought for the first time "die Länder- und Völkerkunde in den Dienst der Weltgeschichtsdarstellung."[195] Helmolt's "great co-operative *History of Mankind* ... emphasizes the sovereign influences of nature and geography," says Gooch.[196]

Rev. H. B. George, in *The Relations of Geography and History*,[197] attempts to "point out systematically how these geographical causes work all history through, first in general, and then in reference to the various countries of Europe,"[198] although "This work does not pretend to attempt the impossible task of describing all the influence exerted by geographical conditions on human history. All that it professes to do is to indicate the modes in which that influence works, with sufficient illustrations from actual history."[199]

Professor Geddes, of Edinburgh, is the most energetic expounder of this idea—the anthropo-geographical conception of history—in the English-speaking world, says Small.[200]

Throughout the entire treatment of Guglielmo Ferrero's[201] *History of Rome* (one of the most original and important historical works of recent years), geography thoroughly permeates history.[202]

Robert Sieger[203] attempts to explain the history and policies of the Austro-Hungarian monarchy "aus ihren geographischen Grundlagen."[204]

Ellsworth Huntington, in *The Pulse of Asia*,[205] illustrates the geographic basis of history.[206]

The Columbia School of sociological historians, and others, interpret history partly in terms of the milieu: physical (economic and geographic) and social.[207]

Human geography, and political geography, have long been divided into fragmentary parts, contended for by economics, history, and sociology.[208] Yet the discipline of anthropo-geography has now become "eine mächtige Hilfswissenschaft der geschichtlichen Auffassung."[209] So that, today, it has become a custom to include in textbooks of history one or more chapters on the relation of geography to history, to show the dependence of history on environment.[210] The study of the latter is a part of Kulturgeschichte or History of Civilization which is defined as embracing the non-political aspects of civilization such as the influence of nature, the pressure of economic factors, the origin and transformation of ideas, the contribution of science and art, religion and philosophy, literature and law, the material conditions of life, the fortunes of the masses.[211] Likewise, only on a broader scale, the milieu is being examined in a new branch of study, which is one resultant of anthropo-geographical research. This new branch of study is economic geography, which, according to John McFarlane,[212] "may be defined as the study of the influence exerted upon the economic activities of man by his physical environment, and more especially by the form and structure of the surface of the land, the climatic conditions which prevail upon it, and the place relations in which its different regions stand to one another." Seligman says that the modern study of economic geography is but an expansion of the study of the influence of milieu.[213]

Indeed, geography itself, *i.e.*, the new geography, is conceived of as the science or study of the responses of organisms to inorganic, and to a certain extent organic, environmental control.[214] Professor William Morris Davis, of Harvard University, is one of the chief exponents of this theory in the United States. Very recently, Rollin D. Salisbury said:[215] "By common consent, Geography (as distinct from physical geography) is the science which deals with the relations of physical environment to life and its activities. In this sense, geography is a connecting link between geology, physiography, and climatology, on the one hand, and zoölogy, botany, sociology, economics, and history on the other. Its subject-matter is in process of formulation...."[216]

More Recent Anthropo-geographical Treatises

James Bryce offers the most excellent general survey of man's relation to his physical environment.[217]

Herbertson's very useful and readable introductory book gives "concrete pictures of human life under these very different conditions typical environments. They show, in the first place, how the occupation of different groups of mankind depends on their geographical surroundings, and how these occupations in turn affect not only the material life, the houses, food, clothing, etc., but also family life, notions of property, progress in trade and manufactures, power of expansion, and ideals of government. All these are classified, not according to race, which is often an accident, but according to those permanent influences by which all races are affected."[218]

Robert DeCourcy Ward, in his standard work on *Climate Considered Especially in Relation to Man*,[219] presents "typical illustrations" of environmental action on the life of man in the tropics (Ch. 8, pp. 220–71), in the temperate zones (Ch. 9 pp. 272–321), and in the polar zones (Ch. 10, pp. 322–37).[220] In a chapter on the hygiene of the zones (Ch. 7, pp. 178–219), Ward also surveys "some of the relations between weather and climate and a few of the more important diseases."[221]

R. R. Marett's chapter on "Environment" in his *Anthropology*[222] presents, beside a number of valuable general and critical remarks, chiefly a regional survey of the world showing the general effect of geographical environment on man.

Camille Vallaux, in *Géographie Sociale, Le Sol et L'État*,[223] beginning with the sixth chapter, also discusses some phases of what would in E. C. Hayes' classification[224] be called the technical milieu.

The most recent German essay, Willy Hellpach's[225] *Die Geopsychischen Erscheinungen: Wetter, Klima und Landschaft in ihrem Einfluß auf das Seelenleben*,[226] deals with the *direct* effects of the surrounding *atmosphere* and soil on the human psyche.[227] Hellpach seems primarily interested in "Psycho-Pathologie";[228] he lays most stress on *das Pathologische*, particularly in the main—first two—parts of his essay: "Wetter und Seelenleben," and "Klima und Seelenleben," where the pathological effect is strongly emphasized. Hellpach's valuable summary of what we know today of this phase of the milieu,[229] revealing as it does by the meager number of the facts assembled the crying need for many more such facts, may be, in its results, somewhat disappointing[230] for the present day, but it augurs well for future investigation.

The latest extensive presentation of general anthropo-geography,[231] Jean Brunhes' *La géographie humaine*,[232] pays more attention to present than to historical conditions,[233] and thus fittingly complements Ellen C. Semple's *Influences of Geographic Environment*,[234] which "may be regarded as superseding Ratzel's great work on Anthropo-geography."[235]

Primitive Peoples and Environment

Karl Ritter, in the essay "Über das historische Element in der geographischen Wissenschaft" (1833), declares that the forces of nature which at the commencement of human history exerted a very decisive influence were bound to recede more and more, and their action had to diminish, in proportion to man's progress. Civilized mankind extricates itself gradually, like single man, from the immediately conditioning fetters of nature and of its place of abode.[236] This opinion of Ritter's was adopted by many.[237]

Theodor Waitz regards primitive man both as purely a product of, and as being completely at the mercy of, circumambient nature: "Denken wir uns vom Menschen Alles hinweg, was an ihm Wirkung der Kultur ist, so steht er da als bloßes Produkt der Macht, die ihn in's Leben rief, ... Das Erste, was an ihm charakteristisch für uns hervorträte, würde die sehr vollständige Abhängigkeit sein, in der er sich von seiner Naturumgebung befände: der gesammte Inhalt, den sein inneres Leben zunächst gewönne, würde ein ziemlich reines Produkt dieser letzteren sein. Der Naturmensch wird zunächst nur das, wozu die Naturverhältnisse ihn machen, unter die er sich gestellt findet; wovon er sich nährt, das werden diese ihm darbieten, auf welche Weise und durch welche Mittel er seine Nahrung gewinnt, dazu werden diese ihm Anleitung geben müssen; ob er Kleidung und sonstigen Schutz gegen äußere Schädlichkeiten bedarf, und wie er diesem Bedürfnis abzuhelfen strebt, werden sie ihn lehren und die Erfindungen, die hierzu nötig sind, ihm an die Hand geben müssen; sie werden mit einem Wort seine ganze Lebenseinrichtung bestimmen ..."[238]

G. Gerland holds that man developed from and upon nature, on which he is very closely dependent and of which he is a small part, and that the higher he rises the more he frees himself from the compelling influence of the earth, which, however, he can never wholly escape.[239]

In the opinion of Herbert Spencer, the earlier stages of social evolution are far more dependent on local conditions than the later stages. They are more at the mercy of their surroundings.[240] Both Spencer and Benjamin Kidd believe that primitive man is at the mercy of the milieu.[241] The "remotely ancient representatives of the human species ... were in their then wild state much more plastic than now to external nature," according to Wallace.[242] Historical and statistical geography show us "die Menschen, wie sie in ihre aktive Rolle eingetreten sind und durch Arbeit die Überlegenheit über das Milieu gewinnen, das sie umgibt ... Nachdem der Mensch ganz den Einfluß des Milieu über sich ergehen ließ, hat er denselben zu seinem Nutzen umgestaltet ..."[243] The intimate connection of first civilizations with physical environment slackens with subsequent advance.[244] This apparently deep-rooted view is controverted by Ratzel who flatly contradicts it. Distinguishing between the direct and the indirect effects of milieu, he argues in straight opposition that with progressing civilization we are increasingly dependent on environment, that the degree of such dependence has not lessened with advancement in civilization, and that only the manner of the relation has changed.[245] Environment affects even the highest civilization, says Ripley.[246] G. Elliot Smith maintains that "Environment, however it may act, whether directly or indirectly, is still helping to shape the human form, and is affecting the development of Man's customs and achievements at least as powerfully as, if not more so than, ever before."[247]

Society and Physical Milieu

The social evolution proceeds amidst the entire system of exterior conditions (chemical, physical, astronomical), by which its rate of progress is determined. Social phenomena can no more be understood apart from their environment than those of individual life.[248] The study of social evolution presupposes a relation to the physical milieu: "Das Studium der sozialen Entwicklung setzt eine Beziehung zwischen der Menschheit, welche den Vorgang vollführt, und der Gesamtheit der äußeren Einflüsse voraus, welche letztere man auch die sogenannte Umgebung heißen könnte."[249]

John Stuart Mill asserts that "All phenomena of society are phenomena of human nature, generated by the action of outward circumstances upon masses of human beings."[250]

To Schäffle, in the analysis of the structure and functions of human society there exist as influential factors the external surroundings, on the one hand, and the active elements of the social body (the individual and the population), on the other; for, as Schäffle emphasizes, not only economics, but all social science must take into consideration not only Society, but also Nature, *i.e.*, the natural fund or stock, designated by soil and climate, of the immediate world-surroundings of the social body as the external sphere embracing societary life, and that, not only as a sum total of free possessions, but also as a multiplicity of free, *i.e.*, unsubjugated resistances.[251]

As "the result of a survey of social organizations, considered as machinery in motion, Hermann Post[252] points out very justly that it is useless to attempt to explain social phenomena on the basis of the psychological activities of individuals, as is too commonly assumed, because all individuals whose conduct we can possibly observe have themselves been educated in some society or other, and presume in all their social acts the assumptions on which that society itself proceeds.... It Post's method is the same method, of course, which had already yielded such remarkable results to Montesquieu, and even to Locke. The point of view is no longer that of a Maine or a McLennan.... It is that of a spectator of human society as a whole.... And its immediate outcome has been to throw into the strongest possible relief the dependence of the form and, still more, of the actual content of all human societies on something which is not in the human mind at all, but is the infinite variety of that external Nature which Society exists to fend off from Man, and also to let Man dominate if he can."[253]

Government, War, Progress, and Climate

James Bryce "has recently clearly set forth the climatic control of government in an essay on 'British Experience in the Government of Colonies' (*Century*, March, 1899, 718–729)."[254] Vallaux, however, is sceptical as to the influence of physical environment upon the State.[255] William Ridgeway avers that political and legal institutions are the result of environment.[256]

Far-reaching and weighty historical consequences "have followed from special conditions of climate or weather. Maguire's 'Outlines of Military Geography' (Cambridge, 1899) contains a chapter on the influence of climate on military operations, but this subject has hitherto received little attention. More recently, Bentley, in a presidential address before the Royal Meteorological Society, London, considered the matter."[257] Still more recently, the relation of climate or weather to war has been scrutinized, among others, by F. Lampe in "Der erdkundliche Unterricht,"[258] by Otto Baschin in "Der Krieg und das Wetter,"[259] and by E. Alt in "Krieg und Witterung."[260]

Hellwald, "the well-known traveller and geographer," compiled his "History of Civilization in its Natural Development" in 1874, according to the findings of which, cultural development is "a natural process, conditioned by race, geography, and climate. Civilisation means the mastering of nature and the taming of man.... Hellwald's standpoint is shared, though less aggressively displayed by Henne-am-Rhyn."[261]

To the late meteorologist Cleveland Abbe, "Everything seems to combine to prove that the existing order of events both material and intellectual has been brought about by a slow process of change, due to the interaction of the atoms and masses that constitute the material world around us."[262]

The great diversity of existent civilizations, declares Auguste Matteuzzi, is due to the diversity of the milieus where they developed. In order to discover why any civilization becomes more heterogeneous and more perfect, one must study the geographic milieu where it evolved. The organic and inorganic milieu of evolving ethnic groups constrains human societies to an incessant process of adaptation, and these societies in their turn react upon the milieu and modify it.[263]

In short, says Auguste Comte, "all human progress, political, moral, or intellectual, is inseparable from material progression, in virtue of the close interconnection which, as we have seen, characterizes the natural course of social phenomena."[264]

That civilization is a result of adaptation to environment, physical as well as political, is the view entertained by Bryce, Strachey, and Geikie.[265]

Climate and Man's Characteristics

There are "certain broad, distinguishing characteristics of man in the temperate and tropical zones, in determining which it is reasonable to believe that climate has played a part. Similarly, there has been a natural tendency to attribute certain differences between northerners and southerners in the temperate zones to a difference in climate.... These national differences are proverbial between northern and southern Germans, French, Spanish, Russians, Italians, Arabs, and other peoples. The influence of climate has likewise been traced in the sad, even pessimistic tone of much of the northern literature, and in the gravity and melancholy of modern northern music, as well as of the older northern folk-songs ... even racial distinctions are more or less directly traceable, in many instances, to climate.... Sir Archibald Geikie, in his *Scottish Reminiscences*, has emphasized the climatic influence in producing the grim character of the Scot...."[266]

Tacitus, in the 29th chapter of the *Germania*, assures us that the soil and climate of the land of the Mattiaci caused them to be more bellicose than their neighbors.[267]

Daudet, "who has written an entire novel ('Numa Roumestan') to depict the great influence of the climate of southern Europe upon conduct, says: 'The Southerner does not love strong drinks; he is intoxicated by nature. Sun and wind distil in him a terrible natural alcohol to whose influence every one born under this sky is subject. Some have only the mild fever which sets their speech and gesture free, redoubles their audacity, makes everything seem rosy-hued, and drives them on to boasting; others live in a blind delirium. And what Southerner has not felt the sudden giving way, the exhaustion of his whole being, that follows an outburst of rage or enthusiasm?'"[268]

Draper "emphasized the important historical consequences of the difference in the characteristics of northerners and southerners in the United States, which he attributed largely to climate, and which found expression in the Civil War.... The Boers in Africa have developed along lines different from those of the Dutch in the United States."[269]

Man's Intellect and Physical Environment

Auguste Comte, who "was very slightly affected by German thought," and who, in early youth, came under the influence of the philosophy that had become prevalent in France before the Revolution, "read the works of most of its leading representatives. He accepted its cardinal principle that 'thought depends on sense, or, more broadly, on the environment.'"[270]

Adolf Bastian worked unceasingly "among the conceptions of the Naturvölker—the 'cryptograms of mankind,' as he called them—..., demonstrating first the surprising uniformity of outlook on the part of the more primitive peoples, and secondly the correlation of differences of conceptions with differences in material surroundings, varying with geographical conditions. This second doctrine he elaborated in his *Zur Lehre von den geographischen Provinzen*, in 1886."[271]

Physiology and statistics "show that most human functions are subject to the influence of heat (Lombroso, 'Pensiero e Meteore,' Milan, 1878). It is to be expected, then, that excessive heat will have its effect upon the human mind."[272]

The physiographer, "... looking back over the history of life upon the earth's surface, ... is forced to the conclusion that its highest estate embodied in the moral and intellectual qualities of man has been, in the main, secured by the geographic variations which have slowly developed through the geological ages."[273]

Benno Erdmann, in his "Gedächtnisrede auf Wilhelm Dilthey," observes that in ripe old age Dilthey in the last of his larger works declared that man finds himself determined by the physical world in which mental occurrences appear only as interpolations.[274]

Religion and Physical Milieu

As physical characteristics "are in the main the result of environment, social institutions and religious ideas are no less the product of that environment.... We might just as well ask the Ethiopian to change his skin as to change radically his social and religious ideas. It has been shown by experience that Christianity can make but little headway amongst many peoples in Africa or Asia, where on the other hand Muhammadanism has made and is steadily making progress, ... This is probably due to the fact that Muhammadanism is a religion evolved ... in latitudes bordering on the aboriginal races of Africa and Asia, and that it is far more akin in its social ideas to those of the Negro or Malay than are those of Christianity, ..."[275]

Ernest Renan "points out that the desert is monotheistic, its uniformity suggesting a belief in the unity of God.... In his *Seas and Skies in Many Latitudes* (London, 1888, pp. 42–43), Abercromby gives two maps, showing respectively the areas of Mohammedanism and the districts in Asia and Africa with a mean annual rainfall of less than ten inches. The maps are strikingly similar. The author adds: 'Whether this distribution of a great creed is the result of chance, or of some deep connection between the tenets of that religion and climatic influences, I can not say;—but still the relation is so remarkable that I have thought it well to bring the matter forward.'"[276]

Climate and Conduct

The "frequent and sudden weather changes of the temperate zones affect man in many ways, as do the larger seasonal changes. The relations between weather and conduct have frequently been investigated. Professor E. G. Dexter has made an extended empirical study of the effects of the weather ... Bertillon has collected data on suicides and seasons in France, ..."[277] Dexter studies empirically by means of statistics—plotting certain curves—the relation between temperature, barometric pressure, humidity, wind, character of the day, precipitation, on the one hand, and the child in school—work, deportment, attendance—, crime, insanity, health—sickness and death—, suicide, drunkenness, attention—errors in calculation made by clerks in banks—, on the other.[278] Of his general conclusions[279] the first is: "Varying meteorological conditions affect directly, though in different ways, the metabolism of life"; the second: "The 'reserve energy' capable of being utilized for intellectual processes and activities other than those of the vital organs is affected *effected*, in the original most by meteorological changes"; the third: "The quality of the emotional state is plainly influenced by the weather states"; the fourth: "Although meteorological conditions affect the emotional states, which without doubt have weight in the determination of conduct in its broadest sense, it would seem that their effects upon that portion of the reserve energy which is available for action are of the greatest import."[280]

The nervous effects of the weather including cyclonic winds have also been noted. Among the Eskimos, "Marriages take place at an early age, especially among the women, and the return of the sun after the long winter has a stimulating effect on the animal passions which leads to sexual excesses of all kinds."[281]

Albert Leffingwell investigates *The Influence of Seasons Upon Conduct*[282] in Great Britain and elsewhere. He formulates the underlying assumption of his inquiry in the following manner: "It is not a new theory, though I propose to carry it somewhat further than it has been pushed hitherto. Over half a century ago, Quetelet in his great work "On Man," suggested the hypothesis.... The hypothesis toward which all the facts point is simply this: that upon the nervous organization of human bodies (perhaps specially upon dwellers in the temperate zones) there is exerted during the procession of the seasons, from winter's close till midsummer, some undefined, specific influence, which in some manner tends to increase the excitability of emotion and passion, and thus also to increase all actions arising therefrom."[283] To mention only one of Leffingwell's illustrations, he brings together in a statistical table the total number of all crimes against persons in England for ten years (1878–87), the same facts for Ireland during the same decade, and for France during forty years (1830–69), and in conjunction therewith says: "Here, again, we find that all crimes, even those arising from personal antipathy or hatred, seem specially prevalent in the warmer half of the year. In England, 55 per cent of all such acts of violence during the ten years 1878–1887 happened in spring and summer, and in France during a period of forty years the average was the same. Ireland, indeed, shows a more even distribution of such crimes; but the tendency is seen even there."[284]

Cesare Lombroso, who is claimed to be the first to have essayed to portray the effect of physical environment on the human psyche,[285] states in his *Criminal Man*,[286] referring to Ferri and Holzendorf, that with high temperature there is an increase in crimes of violence, while low temperature has the effect of increasing the number of crimes against property. In "comparing statistics of criminality in France with those of the variations in temperature, Ferri noted an increase in crimes of violence during the warmer years."[287]

Lombroso, in his *Crime, Its Causes and Remedies*,[288] citing the conclusions of the relevant statistical evidence, establishes that in England and France and Italy the crimes of rape and of murder occur in greatest number in the hottest months; that the maximum number of all rebellions in the whole world between 1791 and 1880 falls everywhere in the hottest month, while its minimum number comes in the coldest months; and that crimes against property markedly increase in the winter.[289]

In the southern parts of Italy and France "there occur many more crimes against persons than in the central and northern portions.... Guerry has shown that crimes against persons are twice as numerous in southern France (4.9) as in central and northern France (2.7 and 2.9). *Vice versa*, crimes against property are more frequent in the north (4.9), than in the central and southern regions (2.3)."[290] According to Buckle,[291] climate makes men's habits regular or irregular.

Climatic Control of Food and Drink

William Ridgeway, summarizing his argument in "The Application of Zoölogical Laws to Man,"[292] says: "We have seen that environment is a powerful factor in the differentiation of the various races of man, alike in physique, institutions, and religion. It is probable that the food supply at hand in each region may be an important element in these variations, whilst the nature of the food and drink preferred there may itself be due in no small degree to climatic conditions.... The aboriginal of the tropics is distinctly a vegetarian, whilst the Eskimo within the arctic circle is practically wholly carnivorous. In each case the taste is almost certainly due to the necessities of their environment.... It is probable that the more northward man advanced the more carnivorous he became in order to support the rigours of the northern climate. The same holds equally true in the case of drink.... All across Northern Europe and Asia there is a universal love of strong drink, which is not the mere outcome of vicious desires, but of climatic law.... This view derives additional support from the well-authenticated fact that one of the chief characteristics of the descendants of British settlers in Australia is their strong teetotalism. This cannot be set down to their having a higher moral standard than their ancestors, but rather, as in the case of Spaniards and Italians (temperance reformers point to the sobriety of the Spaniards, Italians, and other South Europeans), to the circumstance that they live in a country much warmer and drier than the British Isles. We must therefore, no matter how reluctantly, come to the conclusion that no attempt to eradicate this tendency to alcohol in these latitudes can be successful...."[293]

SUMMARY

The Introductory Remark traces the semasiology and use of the word *milieu* and discusses its English and German equivalents "environment" and "Umwelt."

An historical sketch of the milieu idea is then taken up from the very beginnings to the nineteenth century. The earlier notions of environmental influence are general and undifferentiated.

The Hebrew Prophets see the hand of Providence in the harmony of national fate with the configuration of the globe. Hippocrates dwells upon the regularity of climatic effect on man. Aristotle notes the action of physical environment on government and national character. Eratosthenes, Strabo, and other Greek thinkers, relate man causally to surrounding nature. Villani says that the fine air of Arezzo produces great minds. Ibn Khaldūn explains, especially Arabic history, by the circumambient physical and social medium. Michelangelo credits Arezzo's fine air with his mentality. Man is subject to the "skyey influences" hourly (Shakespeare).

Jean Bodin plants the study of environment in French soil so firmly and so successfully that it has since become, in a very real sense, indigenous to France and that Bertillon could justly claim it to be a study "*très-française*," a claim which is true to this very day. Bodin's second contribution is that he undertook, for the first time in the modern period (on the basis of sixteenth century knowledge and experience), a scientific and detailed examination, far-reaching and extensive in scope, of the manifold influences of climatic and geographical conditions upon States, laws, national character, religion, language, temperament, talents and aptitudes,—in brief, upon man's mind, manners, and morals.

The study of milieu thus inaugurated in France by Bodin is set up as a French tradition by Lenglet du Fresnoy, Montesquieu, Turgot,

Cuvier, and others,[294] and has been continued by French writers to our day.

A number of philosophers in the seventeenth and eighteenth centuries take up this idea. The doctrine of environment spreads to England and Germany.

In Germany, Herder becomes the fulcrum of all previous thought (Hebrew, Greek, French, English, and German) on this theory. Herder, in turn, in addition to his other and principal contributions to the theory, affects it by giving a quickened impetus not only to the contemporary development thereof, but also to the later course of that development. Goethe reflects some of Herder's conceptions. Wolf, Niebuhr, the German romanticists—August Wilhelm Schlegel in especial—and Hegel apply Herder's idea to history and continue it therein. Hegel combats the notion that climate can be the be-all and end-all of historical explanation; he implies that climate was held to be a *vera causa*.

The theory of social environment evolves, particularly since Ibn Khaldūn, parallel with that of the physical milieu.

The nineteenth century brings differentiation carried out in human geography including history, in biology, in jurisprudence and economics, in anthropology, in sociology, in literature, and latterly in physics. These disciplines determine our divisions for discussions shortly to follow the present one.

The major portion of this study is then given over to following the milieu idea in some of the more important French, English, and German writers of the past century on what for want of a better name has been called anthropo-geography inclusive of certain aspects of history.

On the whole, their method has been the comparative method. Principles laid down *a priori* would be illustrated by typical cases selected mostly from the past. Or, the process would be reversed to an *a posteriori* reasoning: history restudied to find out its possible connections with the environment. Again: some would pick out a phase of the encompassing medium and follow out its effects in a particular country, while others would try to arrive at a more general conclusion.

With reference to climate in particular, the statistical method was employed by Quételet, Bertillon, Leffingwell, Ferri, Holzendorf, Guerry, Curcio, Lombroso, and others, who established a parallelism, or coincidence, between certain climatic features and the criminal conduct of man.

Delimited aspects of environment, relating again more to climate than any other phase of the milieu, were made the objects of observational or experimentally observational studies by Dexter, Brunhes, and Hellpach, the last two giving the most recent comprehensive summaries of our knowledge in this field. And they are among the best we have.

The next part of this study will continue the survey of the history of this theory in the above mentioned sciences as well as in literature.

APPENDIX

Since the foregoing study was completed, E. Huntington's stimulating book—*vide supra*, p. 79, n.—on *Civilization and Climate* has appeared. He continues what Dexter began. Lack of definiteness in observation, argumentative conviction, reasoned out opinion, are superseded by scientific exactness in ascertaining the action of climate. Chapters 4–7 (pp. 49–147) concern us here. In these chapters he investigates "the exact effect of various climatic factors upon selected groups of people" (p. 49).

Huntington subjects to statistical analysis the daily records of about 550 factory operatives, pieceworkers, employed in three factories in three New England cities. The records, most of them for a complete year, are distributed over the four years from 1910 to 1913 (p. 53).

He computes wage averages. He finds for each working day the average hourly wage for each group of operatives. When the daily averages had been found, they were averaged together by weeks. To give each individual an equal importance, the figures of each group have been reduced to percentages. Finally, the different groups were combined (p. 57). His final computations are represented in curves. A curve, graduated in twelve parts (one for each month), for a given year shows the earnings in percentages at any point and thus reveals the *time* of the weakness or efficiency of the worker; it shows the time of his wages from least to most, thereby indicating the time of his work and energy from poorest to best.

Huntington worked up similarly the records of 65 operatives in a North Carolina factory, of 240 operatives in four cotton mills in South Carolina and Georgia, of 57 carpenters at Jacksonville, Fla., and on a different basis the work of 2700 cigar makers in two cigar factories in Florida. On the first basis he also computed a series of data from a large factory at Pittsburgh, Pennsylvania, based on the work of about 950 operatives in 1910, of about 750 in 1911, of 69 in 1912, of about 7000 in 1913. He figured the monthly or bi-weekly averages of hourly earnings of these pieceworkers in Pittsburgh.

Discussing the curves in Figure 1 (p. 59), he mentions (p. 61) five features revealed by the curves that show no sign of disappearing. They are: "an extremely low place in midwinter, and a less pronounced low place in midsummer; a high point in June, a still higher point at the end of October, and a hump in mid-December....

"Before we discuss the causes of the variability of the summers let us consider the meaning of the curves as a whole. In the first place, it is evident that, although details may vary from year to year, the general course of events is uniformly from low in the winter to high in the fall with a drop of more or less magnitude in summer. To what can this be due?...

"We seem forced to search outside of the factories for the reasons for our seasonal fluctuations of wages.... There seems to be no recourse except to ascribe the fluctuations of the curves to climate pp. 64–5.

"The verity of the conclusion just reached is strongly confirmed by comparison with other regions and other types of human activity.... The curves in Figure 2, pp. 66–7 range from the Adirondacks in northern New York to Tampa in southern Florida and include one from Denmark. With them I have repeated some of the curves of Figure 1 for the sake of comparison. The most remarkable feature of this series is that although there is great diversity of place and of activity, all the curves harmonize with what would be expected on the basis of Figure 1 p. 65.

"The general form of the curves for Pittsburgh and Connecticut is obviously the same....

"The agreement between the curves for Connecticut and Pennsylvania is far too close to be accidental p. 76.

"We have now seen that from New England to Florida physical strength and health vary in accordance with the seasons. Extremes seem to produce the same effect everywhere. The next question is whether mental activity varies the same way" (p. 77).

Huntington uses the marks of "about 1900 students for a single year" in mathematics (weekly averages at Annapolis and daily averages at West Point) and in English (at Annapolis). From these data he compiles the curves in Figure 3 (p. 80). He says (p. 81), "The curves of mental activity all resemble it the average curve of physical work in having two main maxima, in fall and spring.... At Annapolis, just as at West Point, the time of best work is when the mean temperature is not far from forty degrees Fahrenheit.

"Summing up the matter, we find that the results of investigations in Denmark, Japan, Connecticut, Pennsylvania, New York, Maryland, the Carolinas, Georgia, and Florida are in harmony. They all show that except in Florida neither the winter nor the summer is the most favorable season. Both physical and mental activity reach pronounced maxima in the spring and fall, with minima in midwinter and midsummer. The consistency of our results is of great importance. It leads to the belief that in all parts of the world the climate is exercising an influence which can readily be measured, and can be subjected to statistical analysis" (p. 82).

This is his conclusion in Chapter IV (pp. 49–82), "The Effect of the Seasons."

Having seen in the fourth chapter "that both physical and mental energy vary from season to season according to well-defined laws," Huntington investigates in the fifth chapter ("The Effect of Humidity and Temperature," pp. 83–110) "the special features of seasonal change which are most effective" (p. 83). Explaining the curves of Human Activity and Mean Temperature (p. 99), he says (p. 98), "With the exception of the last two, which are distinctly the least reliable, the physical group all reach maxima at a temperature between 59° and 65°. Even the two less reliable curves reach their maxima within the next four degrees. All the curves decline at low temperatures, ..., and also at high.

"Another point brought out by the curves on p. 99 is that as we go to more southerly climes the optimum temperature of the human race becomes higher. It is important to note, however, that the variation in the optimum is slight compared with the variation in the mean temperature of the places in question. For instance, in Connecticut the optimum seems to be about 60° for people of north European stock. This is about ten degrees higher than the mean temperature for the year as a whole. In Florida, on the other hand, the optimum for Cubans is about 65°, which is five degrees *lower* than the mean temperature for the year at Tampa. In other words, with a difference of twenty degrees in the mean annual temperature, and with a distinctly northern race compared with a southern, we find that the optimum differs only about 5° F. This seems to mean that for the entire human race the optimum temperature probably does not vary more than ten or fifteen degrees pp. 100–101.

"The last thing to be considered in Figure 8 p. 99 is the mental curve showing optimum mental work at 38° F. at the bottom. It is based on so large a number of people, and is so regular, that its general reliability seems great, although I think that future studies may show the optimum to be a few degrees higher than is here indicated. It agrees with the results of Lehmann and Pedersen. Furthermore, from general observation we are most of us aware that we are mentally more active in comparatively cool weather. Perhaps 'spring fever' is a mental state far more than a physical. Apparently people do the best mental work on days when the thermometer ranges from freezing to about 50°—that is, when the mean temperature is not far from 40°. Inasmuch as human progress depends upon a coördination of mental and physical activity, we seem to be justified in the conclusion that the greatest total efficiency occurs halfway between the mental and physical optima, that is, with a mean temperature of about 50°" (pp. 102–103).

The curves (p. 105) on Mean Temperature and Vital Processes in Plants, Animals and Man show physical energy to be at the optimum at the mean temperature of 60° F., mental energy at 38°, mental and physical energy combined at from 40° to 60°. Of this last mentioned curve he says: "It may be taken as representing man's actual productive activity in the things that make for a high civilization. The resemblance of the human curves to those of the lower organisms is obvious. In general, the lower types of life, or the lower forms of activity, seem to reach their optima at higher temperatures than do the more advanced types and the more lofty functions such as mentality. The whole trend of biological thought is toward the conclusion that the same laws apply to all forms of life. They differ in application, but not in principle. The law of optimum temperature apparently controls the phenomena of life from the lowest activities of protoplasm to the highest activities of the human intellect" (pp. 109–110).

In Chapter VI ("Work and Weather," pp. 111–128), he interprets the curves he plotted showing especially the influence of changes of temperature from day to day, and of the character of each day and its relation to storms. In the very interesting Chapter VII (pp. 129–147) he discusses "The Ideal Climate."

In the closing paragraph of his book, he says, "If our hypothesis is true, man is more closely dependent upon nature than he has realized. A realization of his limitations, however, is the first step toward freedom p. 293.

"The hypothesis, briefly stated, is this: Today a certain peculiar type of climate prevails wherever civilization is high. In the past the same type seems to have prevailed wherever a great civilization arose. Therefore, such a climate seems to be a necessary condition of great progress. It is not the cause of civilization, for that lies infinitely deeper. Nor is it the only, or the most important condition. It is merely one of several, ..." (p. 9.)

Huntington mentions (p. 7) Lehmann and Pedersen's "Das Wetter und unsere Arbeit" and Berliner's "Einfluß von Klima, Wetter und Jahreßeit auf das Nerven- und Seelenleben," without the date or place of publication.

NOTE: Since the foregoing pages went to press, the following publications have appeared; being too late for inclusion or comment in the text, they are added here for reference:

Douglas W. Johnson, *Topography and Strategy in the War*, N. Y., Henry Holt & Co., 1917, 221 pp. (Thorough and very illuminating; points out how the surface features of the country influenced military operations in the most important theaters of the war.)

James Fairgrieve, *Geography and World Power*, N. Y., E. P. Dutton & Co., 1917, 356 pp. (Shows how History has been controlled by Geography.)

Robert De C. Ward, "Weather Controls Over the Fighting in the Italian War Zone," *The Scientific Monthly*, Vol. 6, No. 2 (February, 1918), pp. 97–105. And "Weather Controls Over the Fighting in Mesopotamia, in Palestine, and near the Suez Canal," *ibidem*, Vol. 6, No. 4 (April, 1918), pp. 289–304.

1. For brief but valuable sketches of one phase or another of the history of the theory of milieu, cf. Friedrich Ratzel, *Anthropogeographie*. 1. *Teil: Grundzüge der Anwendung der Erdkunde auf die Geschichte* (2. Aufl., Stuttgart, 1899, 604 pp.), pp. 13–23, 25–30, 31–40; Gustav Schmoller, *Grundriß der Allgemeinen Volkswirtschaftslehre*. Erster Teil (Vierte bis sechste Aufl., Leipzig, 1901), p. 127, pp. 137 f., 144 ff., Zweiter Teil (Erste bis sechste Aufl., Leipzig, 1904), pp. 656 ff.; *Ferdinand v. Richthofen's Vorlesungen über Allgemeine Siedlungs- und Verkehrsgeographie*, bearb. und herausgegeben von O. Schlüter (Berlin, 1908, 351 pp.—A course of lectures delivered in the summer semester of 1891 in Berlin, repeated in the winter semester in 1897/8), pp. 6–13; Jean Brunhes, *La Géographie Humaine* (Deuxième édition, Paris: Alcan, 1912, 801 pp.), pp. 36 ff.; A. C. Haddon and A. H. Quiggin, *History of Anthropology* (London, 1910, 158 pp.), pp. 131 f., 150–52; William Z. Ripley, "Geography and Sociology," *Political Science Quarterly*, X (1895), pp. 636–54; also the same author's *The Races of Europe* (New York: D. Appleton & Co., 1899), pp. 2–5. Cf. also O. Schlüter, "Die leitenden Gesichtspunkte der Anthropogeographie, insbesondere

der Lehre Friedrich Ratzels," *Arch. f. Sozialwissenschaft*, Bd. IV (1906), S. 581–630, and Rudolf Goldscheid, *Höherentwicklung und Menschenökonomie*, I Philosophisch-soziologische Bücherei, Band VIII, (Leipzig: W. Klinkhardt, 1911, 664 pp.), p. 52. For bibliographies, in addition to those yet to be mentioned, see also Ratzel, *l.c.*, pp. 579–85; Brunhes, *l.c.*, nn.; Ellen C. Semple, *Influences of Geographic Environment, On the Basis of Ratzel's System of Anthropo-geography* (New York: H. Holt & Co., 1911, 637 pp.), to each chapter of which an extensive bibliography is added; William J. Thomas, *Source Book for Social Origins* (Chicago and London, 1909) pp. 134–39: Bibliography to Part I: The Relation of Society to Geographic and Economic Environment (pp. 29–129, Comment on Part I, pp. 130–33); Ripley, "Geography and Sociology," *Pol. Sc. Quar.*, X (1895), pp. 654–5.

2. *Dictionnaire de l'Académie Françoise.* Quatrième Édition. Tome Second (Paris, 1762), p. 143.

3. *Encyclopédie, ou Dictionnaire Raisonné des Sciences*, etc. Nouvelle Éd. 1778, ed. by Diderot and D'Alembert, 21st vol., p. 853.

4. *Cours de Philosophie Positive* (6 vols., 1830–42, 5ᵉ édition, Paris, 1892–94), see vol. 3, p. 235 n.

5. Cp. esp. the Introduction to his *Histoire de la Littérature Anglaise*, 5 Tomes (8ᵉ Édition, Paris: Hachette, 1892); the first edition appeared in 1863, after Taine had been at work on it for well-nigh a decade.

6. For Zola as the disciple of Taine, cf. H. Wiegler, *Geschichte und Kritik der Theorie des Milieus bei Émile Zola* (Diss., Rostock, 1905), esp. pp. 19–36.

7. Vide Émile Waxweiler, *Esquisse d'une Sociologie* (Bruxelles, 1906), p. 65.

8. *Dictionnaire de la Langue Française*, vol. 3 (1885), pp. 559 f.

9. *Verdeutschungen, Wörterbuch fürs tägliche Leben* (Braunschweig, Verlag von George Westermann, 1915, 176 pp.), p. 93.

10. *Verdeutschungsbücher des Allgemeinen Deutschen Sprachvereins, III* (Zweite Aufl., neu bearb. v. Edward Lohmeyer, Berlin, Verlag des Allgemeinen Deutschen Sprachvereins, 1915, 182 pp.), pp. 91 f.

11. *Phénomènes de la vie* (2ᵉ éd., Paris, 1885), t. I, p. 112. See Waxweiler, *l.c.*, p. 36.

12. *Race Prejudice*, transl. by Florence Wade-Evans (London, 1906), p. 130.

13. "The Services of Naturalism to Life and Literature. Reprinted, with Additions, from *The Sewanee Review*, October, 1903," p. 2.

14. See Murray's NED., vol. III, Part II, (1897), p. 231.

15. *Wörterbuch d. d. Sprache* (1811), Bd. 5, S. 113.

16. See the article by I. Stosch on "Umwelt-*milieu*," *Zeitschrift für Deutsche Wortforschung*, g. v. Fr. Kluge, 7. Bd. (1905), pp. 58–9.

17. 2. Bd., 2. Hälfte (Leipzig: Otto Wigand, 1865), p. 1556ᵇ.

18. A. Gombert cites the passage in question in his article "Umwelt," *Z. f. D. Wf.*, 7. Bd. (1905), pp. 150–52.

19. The Belgian sociologist De Greef, in his *Introduction à la Sociologie* (1886–89), raised "Mésologie" (denoting "Erkenntnis der milieux") to a special introductory branch of sociology for the purpose of discussing, according to Ratzel superficially, the external factors of history; cf. Paul Barth, *Die Philosophie der Geschichte als Soziologie*, I (Leipzig: Reisland, 1897), p. 70 and Ratzel, *l.c.* p. 29. The term "Mésologie" was in use in France at an earlier date than that. See for example the title of an article written at the close of the Franco-German war by Dr. Bertillon, "De l'Influence du milieu ou Mésologie," *La Philosophie Positive*, Revue dirigée par É. Littré & G.

Wyrouboff, Tome IX (Paris, 1872), pp. 309–20. Or see M. E. Jourdy, "De l'Influence du milieu ou Mésologie," *ibid.*, Tome X (1873), pp. 154–60.

20. Fr. de Rougemont, in his important work *Les deux cités; la philosophie de l'histoire aux différents âges de l'humanité* (1874) treats this question exhaustively. See Robert Poehlmann, *Hellenische Anschauungen über den Zusammenhang zwischen Natur und Geschichte* (Leipzig: S. Hirzel, 1879, 93 pp.), pp. 8 f.

21. *Vide* Eugénie Dutoit, *Die Theorie des Milieu* (Diss., Bern, 1899, 136 pp.), pp. 52–5.

22. "Hippocrate fut le premier à observer quelques-uns des effets du milieu sur l'individu. Ses observations sont nécessairement nébuleuses et chaotiques, plutôt descriptives et qualitatives, étant donnée l'imperfection des connaissances de son temps."—Auguste Matteuzzi, *Les Facteurs de l'Évolution des Peuples* (Paris, 1900), p. 6 (Avant-Propos).

23. "Wir sahen, daß sich das Buch des Hippokrates durchaus darauf beschränkte, die Wechselbeziehungen zwischen Landesnatur und Volkscharakter zu erörtern."—Poehlmann, *l.c.*, p. 51.

24. "Hippokrates von Kos, 'der Vater der Heilkunde' (ca. 460 bis ca. 370), ist der *Begründer der Anthropogeographie*. Er schrieb ein Buch über Klima, Wasser und Bodenbeschaffenheit und ihren Einfluß auf die Bewohner eines Landes in physischer und geistiger Beziehung. Der philosophische Gedanke war damit angeregt, fand aber keine weitere Entwicklung."—*F. v. Richthofen's Vorlesungen*, etc. (Berlin, 1908), p. 7.

25. *System of Positive Polity* (4 vols., London: Longmans, Green & Co., 1875–77—the original was published in 1851–54), vol. II, p. 364: "... a study of the aggregate of material influences: Astronomical, Physical, Chemical which was commenced by the great Hippocrates in his admirable and unequalled Treatise upon Climate."

26. Haddon and Quiggin, *Hist. of Anthropology* (1910), p. 150.—Poehlmann discusses Hippocrates in *Hellenische Anschauungen*, etc., pp. 12–37.—Ludwig Stein, in his book *Die soziale Frage im Lichte der Philosophie* (2. verb. Aufl., Stuttgart, 1903), p. 403, n., says that "Aless. Chiapelli, *Le promesse filosofiche del Socialismo* (Napoli, 1897), p. 41, hebt die interessante Tatsache hervor, daß die Lehre vom 'Milieu' ihrem Keime nach auf Hippokrates zurückgeht." But a little over three decades earlier, Peschel in his *Geschichte der Erdkunde* (1. Aufl., 1865) surveyed on two pages some important phases of Hippocrates and Strabo on milieu. And earlier still, a half century before Peschel, Ukert in his *Geographie der Griechen und Römer* (1816), I, 1, 79, noted Hippocrates as carefully observing the effect of climate on the body and mind of man. (*Vide* Poehlmann, l.c., pp. 7 f.)—And to Herder, Hippocrates was the principal author on climate: "... *Hippocrat. de aere, locis et aquis*, ... Für mich der Hauptschriftsteller über das Klima."—*Herders Sämmtliche Werke*, hg. v. B. Suphan, 13, 269 n.

27. See Dutoit, *Die Theorie des Milieu*, pp. 55–8.

28. Poehlmann, *l.c.*, p. 68.—Aristotle neglects to give credit to Hippocrates in connection with his ideas on environment, although indebted to Hippocrates whom he mentions elsewhere. See Dutoit, *l.c.*, p. 57.

29. "Varron, *De re rustica*, 1, cite une oeuvre d'Eratosthènes où celui-ci cherchait à démontrer que le caractère de l'homme et la forme du gouvernement sont subordonnés au voisinage ou à l'éloignement du soleil. Tentative sublime mais prématurée, pour ramener les phénomènes sociaux à des lois uniques et générales."—Auguste Matteuzzi, *Les Facteurs de l'Évolution des Peuples* (Paris, 1900), p. 6.

30. "Die vollständigste Beschreibung of the earth gab erst Strabo in seinem Werk γεωγραφικά. Hier begegnen wir zum zweitenmal der philosophischen Idee, *Mensch und Natur in Kausalzusammenhang* miteinander zu bringen. Strabos Geographie ist als 'Länder- und Völkerkunde' das größte Werk des Altertums. Die Anschauung eines kausalen Zusammenhanges des Menschen mit der Natur ging darauf unter according to him, until the middle of the eighteenth century, until Montesquieu."—*Richthofen's Vorlesungen*, etc. (1908), p. 8.

31. Buckle and his Critics (London, 1895, 548 pp.), p. 7 n.

32. See Poehlmann, *l.c.*, p. 7.—For a brief statement of the theory of milieu in Greek writers (Thucydides, Xenophon, Plato, Aristotle, Theophrastus), cf. Curtius, *Boden und Clima von Athen* (1877), p. 4 f. For Aristotle, compare also Dondorff, *Das hellenische Land als Schauplatz der althellenischen Geschichte* (Hamburg, 1899, 42 pp.), pp. 11 f. Poehlmann, *l.c.*, discusses the views on environment of Herodotus (pp. 37–47), of Thucydides (pp. 52–4), of Xenophon (pp. 55 f.), of Ephoros only fragments of his great work, A Universal History, are extant; cited by Strabo (pp. 56–9), of Plato (pp. 59–64), of Aristotle (pp. 64–74), of Polybios (pp. 75–7), of Posidonios in Strabo and in Galen (pp. 78–80), of Strabo (pp. 80–90), of Galen (pp. 91 f.).

33. Vide Élisàr v. Kupffer, *Klima und Dichtung, Ein Beitrag zur Psychophysik* in *Grenzfragen der Literatur und Medizin* in Einzeldarstellungen hg. v. S. Rahmer, Berlin, 4. Heft (München, 1907), p. 63.

34. Translated into French by Baron Meg. F. de Slane (3 vols., Paris, 1862-8).

35. See R. Flint, *History of the Philosophy of History, Historical Philosophy in France and French Belgium and Switzerland* (New York: Scribner, 1894, 706 pp.), pp. 159 f.—"His Mohammed Ibn Khaldūn's fame rests securely ... on his *magnum opus*, the 'Universal History,' and especially on the first part of it, the 'Prolegomena' (p. 162).... They the Prolegomena may fairly be regarded as forming a distinct and complete work.... It consists of a preface, an introduction, and six sections or divisions (p. 163)."

36. Flint, *l.c.*, pp. 164 f.

37. *Vide infra*, p. 27.

38. Flint, *l.c.*, p. 164.—Cf. also pp. 158–72, for Ibn Khaldūn in general.

39. Cf. Kupffer, *Klima and Dichtung*, p. 63.

40. "Da Bodin hauptsächlich an die Anschauungen des Aristoteles anknüpft, ...—Auch an Strabo, der dem Einfluß des Klimas und der Landesnatur schon die schöpferischen Kräfte des Volksgeistes gegenübergestellt hat, lehnt sich Bodin an."—Fritz Renz, *Jean Bodin, Ein Beitrag z. Geschichte d. hist. Methode im 16. Jahrhundert* Geschichtliche Untersuchungen hg. v. Karl Lamprecht, III. Bd., I. Heft, (Gotha, 1905, 84 pp.), p. 48 n.

41. *Methodus ad facilem historiarum cognitionem*, published in 1566.

42. Flint, *l.c.*, 198.—The 'Republic' was first published in 1576 in French under the title *De la République*. Eight years later (1584) Bodin himself translated it into Latin as *De Republica Libri Sex*. See Ludwig Stein, *Die soziale Frage im Lichte der Philosophie* (2. verb. Aufl., Stuttgart, 1902), p. 217 n.

43. Compare Dutoit, *Die Theorie des Milieu*, pp. 58–62.

44. "Die physische Konstitution des Menschen hängt nach Bodin eng mit den klimatischen Verhältnissen seiner Heimat zusammen und entspricht dem Verhalten der Erde, die er bewohnt ..."—Renz, *Jean Bodin* (1905), p. 50.—"... Da der animalische Körper wie alle Körper aus einer Mischung der Elemente besteht, so ergibt sich eine direkte Abhängigkeit der physischen Konstitution von der umgebenden Natur, ja sogar eine Übereinstimmung mit dem Verhalten der Erde in dem betreffenden Himmelsstrich. Der menschliche Körper reagiert auf die klimatischen Einflüsse genau so wie die Erde, die er bewohnt, ..."—*Ibidem*, p. 44.

45. Discussed by Renz, *l.c.*, pp. 47–61, in the chapter "Die Theorie des Klimas."—"Behandelt wird die Theorie des Klimas nach dem 5. Kapitel des 'Methodus,' in dem sich Bodin zum ersten Male mit dieser Doktrin befaßte; zur Erläuterung wird auch das 1. Kapitel des V. Buches der 'République' herangezogen, in dem die Theorie des Klimas, aber in gedrängterer Form, wiederholt wird."—*Ibid.*, p. 47 n. Cf. also p. 45.

46. "Sogar das Temperament variiert nach dem Klima ...

"Wie das Temperament wird die Sprache von dem inneren physischen Bau abhängig gedacht ...

"Ebenso wird die Fortpflanzungsfähigkeit in direkte Abhängigkeit von der physischen Konstitution gebracht ..."—*Ibid.*, pp. 52 f.

47. "Wie das Äußere und die physische Konstitution hängen auch die Anlagen und Fähigkeiten der Völker mit den klimatischen Verschiedenheiten zusammen ..."—*Ibid.*, p. 54.

48. "... Nach der Dreiteilung der seelischen Fähigkeiten bei dem Einzelmenschen und den Bewohnern jedes Staates werden die Völker auf der ganzen Erde gruppiert, indem durch das Klima immer eine Anlage besonders zur Ausbildung kommt ..."—*Ibid.*, p. 46.

49. "... Bodin nimmt zwei Teile des menschlichen Seelenlebens an, erstens eine allen Menschen gemeinsame, unveränderliche geistige Befähigung, die Vernunft, und zweitens Anlagen, die von dem Klima und der physischen Natur des Menschen abhängen. In der 'République' wird ausgeführt, daß diese abhängigen Anlagen nur verschiedene von dem geographischen Milieu abhängige Entwicklungsstufen des Verstandes sind, während dieser an sich von den einzelnen Gegenden unabhängig ist ..."—*Ibid.*, p. 45.

50. "... Indem er Bodin als erster in der Neuzeit auf streng wissenschaftlicher Grundlage versucht, die Wechselwirkung, die zwischen dem historischen Verlauf und der Natur stattfindet, festzustellen, gelangt er zu der Annahme von zwei Teilen des geistig-

seelischen Innenlebens, eines von den umgebenden Verhältnissen abhängigen und eines absoluten, gegen äußere Einflüsse sich passiv verhaltenden Teils. Willensfreiheit neben der durch das Milieu bedingten Ausbildung bestimmter Anlagen und Fähigkeiten ist der mittlere Weg, den er zwischen der Annahme des zwingenden Einflusses der äußeren Natur und der gänzlichen Unabhängigkeit von ihr einschlägt ..."—*Ibid.*, p. 77.

51. "Neben dem Horizontal- wendet Bodin den Vertikalmaßstab zur Beurteilung der Völker an, indem er untersucht, wie die verschiedene Erhebung des Bodens auf die Gestaltung des Volkscharakters einwirkt ...

"Ebenso wird die Natur der Völker von der Qualität des heimatlichen Bodens beeinflußt, ..."—*Ibid.*, p. 58.—"Der Einfluß, der sich aus der östlicheren oder westlicheren Wohnlage auf den Volkscharakter geltend macht, ist, wo nicht in der Richtung Süd-Nord sich erstreckende Gebirge eine deutlichere Scheidelinie bilden, nach Bodin schwer zu bestimmen ..."—*Ibid.* p. 57.

52. "Neben der Vorstellung von der geistig-sittlichen Einheit der Menschen geht die Erkenntnis der Verschiedenartigkeit der Nationen und ihres Bildungsgrades her, die aus den partikularen Bedingungen des nationalen Einzeldaseins resultiert. Zur Erklärung des Volkscharakters wird, wie schon dargelegt, die Theorie des Klimas herangezogen ..."—*Ibid.*, p. 62.

53. "Bodin hat sich deswegen mit der Theorie des Klimas beschäftigt, weil er in der Geschichte und im Völkerleben bestimmte regelmäßige Erscheinungen wahrnahm, die er sich nur aus dem Einfluß des geographischen Milieus erklären konnte. Bei dem strengen Festhalten an der menschlichen Willensfreiheit konnte er sich diesen Einfluß nur durch die Annahme einer von äußeren Verhältnissen abhängigen Entwicklungsfähigkeit der geistigen Anlagen in bestimmter Richtung erklären..."—*Ibid.*, p. 60 f.—"Das unbedingte Festhalten an der menschlichen Willensfreiheit mußte Bodin vor der Annahme bewahren, daß der Einfluß des geographischen Milieus auf die Menschen ein zwingender sei. Nur die Entwicklung der Anlagen wird von der Umwelt bestimmt, nicht aber das sittliche Wollen ..."—*Ibid.*, p. 59.

54. "Wo die äußere Natur zur Entwicklung schlechter Anlagen führt, besitzt nach Bodin die Menschheit in der Erziehung ein Mittel, diesem Übelstand zu begegnen."—*Ibid.*, p. 77.—"... den Menschen wird die Fähigkeit zugesprochen ..., die schädlichen Einwirkungen des Klimas wenn auch schwer, zu überwinden ..."—*Ibid.*, p. 60.

55. *L.c.*, p. 198.

56. "... Den Vergleich der drei Völkergruppen südliche, mittlere, nördliche mit den menschlichen Lebensaltern hat Bodin von Aristoteles entlehnt, was er Meth. V 140, 141 selbst zugibt."—Renz, *l.c.*, p. 57.

57. *L.c.*, p. 48.

58. Haddon and Quiggin, *Hist. of Anthropology* (London, 1910), p. 150.

59. *L.c.*, p. 77.—For Bodin in general, cf. Renz, *Jean Bodin*; Flint, *l.c.*, pp. 190–200; Ludwig Stein, *Die soziale Frage im Lichte der Philosophie*, pp. 217–19. H. Morf, *Französische Literatur im Zeitalter der Renaissance* (2. verb. Aufl., Straßburg: Trübner, 1914), is brief on Bodin, *vide* esp. pp. 131 f.; cf. also p. 125.

60. *Vide* E. Bernheim, *Lehrbuch der historischen Methode* (5. u. 6. Aufl, Leipzig, 1908), p. 230.

61. Montesquieu, *The Spirit of Laws* (translated from the French by Th. Nugent, new ed., revised by J. V. Prichard, 2 vols., London: Geo. Bell and Sons, 1906), I, 238–314.

62. "Seine Montesquieu's Hervorkehrung des Einflusses, den Klima und Bodenbeschaffenheit auf die Soziabilität der Menschennatur

ausüben, geht ebenfalls auf Locke, weiterhin auf Bodin zurück."—L. Stein, *Die soziale Frage*, etc., p. 364.—According to Dutoit (*Die Theorie des Milieu*, p. 62), Montesquieu concealed his obligation to Bodin.

63. *L.c.*, pp. 238–53.

64. *L.c.*, pp. 253–69.

65. *L.c.*, pp. 270–83.

66. *L.c.*, pp. 284–91.

67. *L.c.*, pp. 291–314.

68. Flint, *l.c.*, pp. 279 f.

69. Flint, *l.c.*, p. 286.—(Turgot died in 1781.)

70. Ripley, *The Races of Europe* (1899), p. 4.—Cuvier was twenty years younger than Goethe; both died in the same year.

71. E. G. Conklin, *Heredity and Environment in the Development of Men* (Princeton Univ. Press, 1915, 533 pp.), p. 303.

72. *Eckermanns Gespräche mit Goethe*, neu herausgegeben v. H. H. Houben (Leipzig: Brockhaus, 1909), p. 264.

73. *Ibid.*, p. 265.—These two passages are also cited by Kupffer, *Klima and Dichtung*, p. 64.

74. *Eckermanns Gespräche mit Goethe*, p. 542.

75. *Ibid.*, p. 546.

76. Karl Lamprecht, "Neue Kulturgeschichte" (pp. 449–64 in *Das Jahr 1913, Ein Gesamtbild der Kulturentwicklung*, hg. v. D. Sarason, Leipzig-Berlin: B. G. Teubner, 1913), p. 453.

77. Albert Poetzsch, *Studien zur frühromantischen Politik und Geschichtsauffassung* (Leipzig: Voigtländer, 1907, 111 pp.), p. 89.

78. "Die Einwirkung der äußeren Natur auf die Geschichte tritt zurück in der romantischen Geschichtsphilosophie"; and in a note is added: "Wenn auch der Zusammenhang von Boden und Geschichte, namentlich von natürl. Grenzen u. Staat, der Betrachtung nicht verloren geht. Vgl. A. W. Schlegel, Enz. 216. 697."—*Ibid.*, p. 94.

79. Bernheim, *Lehrb. d. hist. Methode*, p. 650.

80. *Ibid.*, p. 515.

81. See Ludwig Gumplowicz, *Der Rassenkampf* (2.... Aufl., Innsbruck, 1909), p. 9 n.

82. *Vide* the quotation from Hegel by Gumplowicz, *l.c.*, p. 13 n.

83. This paper will carry the discussion through anthropo-geography.

84. The whole question, including Herder's own idea thereof and his indebtedness to preceding authors, both German and foreign, as well as his influence upon succeeding writers at home and abroad, his relation to his contemporaries, etc., will be essayed more fully in a series of papers, to be published soon, dealing with "Herder's Conception of Milieu," "Herder's Relations to France," "Herder's Relations to England," and "Herder in His Own Milieu."

85. The term "anthropo-geography" derives from the title of Fr. Ratzel's main work.—"... le domaine si intéressant, mais à peine défriché, de l'*anthropogéographie*, semble avoir acquis à ce mot le droit de cité dans le langage scientifique."—L. Metchnikoff, *La Civilisation et Les Grands Fleuves Historiques* (Paris, 1889), p. 70 and n.—In England, and in America, it is commonly called human geography, after the French "la géographie humaine." Various names have been proposed for this subject. See also W. Z. Ripley,

"Geography and Sociology." The Viennese Erwin Hanslick, I believe, denominates it "Kulturgeographie."

86. Walther May, "Herders Anschauung der organischen Natur," *Archiv f. d. Geschichte der Naturwissenschaften u. d. Technik*, etc., Leipzig, Bd. 4 (1913, S. 8–39, 89–113), p. 91.

87. *Ferd. v. Richthofen's Vorlesungen üb. Allgem. Siedlungs- u. Verkehrsgeographie*, bearb. u. hg. v. O. Schlüter (Berlin, 1908), p. 11.

88. "... Ritter selbst hat keine methodische Darstellung, kein Lehrgebäude gegeben; sondern nur Andeutungen, die anregend sind. Daher blieb Ritters Grundidee fast ohne Einfluß auf die Geographie; nur die Historiker haben sie sich angeeignet und haben seitdem größeres Gewicht auf die Landesnatur gelegt."—*Ibid.*, p. 11.

89. *Cosmos, a Sketch of a Physical Description of the Universe*, translated by E. C. Otté (5 vols., New York: Harper, 1875–77), p. 48.

90. *Die Erdkunde im Verhältnis zur Natur und zur Geschichte des Menschen oder eine allgemeine, vergleichende Geographie* was published in two volumes at Berlin in 1817–18; the second edition, completely revised, appeared in nineteen volumes from 1822 to 1859, the year of his death. Neither edition is finished; the second deals only with Africa (vol. 1) and Asia (vols. 2–19).

91. *Die Naturkunde*, etc.—See Th. Achelis, *Moderne Völkerkunde* (Stuttgart, 1896), p. 71.

92. *Ibid.*, see Achelis, *l.c.*, pp. 72 f.

93. In Felix Lampe's book, *Große Geographen, Bilder aus der Geschichte der Erdkunde* (Leipzig u. Berlin: B. G. Teubner, 1915, 288 S. Band 28 der v. B. Schmid in Zwickau herausgegebenen "Naturwissenschaftlichen Bibliothek"), neither the chapter on Ritter (pp. 227–33), nor that on "Die wissenschaftliche Geographie der Gegenwart" (pp. 281–87) is very full.

94. Stuttgart & Tübingen, 1808.

95. *Views of Nature* (London, 1850), Author's Preface, p. X.

96. p. 382. See Achelis, *Moderne Völkerkunde*, pp. 88 f.—The relation of man to environment is also referred to in *Cosmos* (English translation by Otté), I, pp. 351-9.—*Kosmos* was originally published as follows: vols. 1 and 2 in 1845-7; vols. 3 and 4 in 1850-8; vol. 5 in 1862.

97. Leipzig, 1841.

98. Kohl, *Der Verkehr*, etc., p. 111. See Achelis, *l.c.*, pp. 80 f.

99. Ibid.

100. Kohl, *l.c.*, p. 537. See Achelis, *l.c.*, pp. 81 f.

101. Kohl, *Ibid.*,—See Achelis, *l.c.*, pp. 82 f.—The manifold influences of nature are also exemplified in Kohl's *Die geographische Lage der Hauptstädte Europas*, 1874, and L. Felix, *Der Einfluß der Natur auf die Entwicklung des Eigentums*, 1893.

102. *Über den Einfluß der äußeren Natur auf die sozialen Verhältnisse der einzelnen Völker und die Geschichte der Menschheit überhaupt, 1848*; later published in *Studien aus dem Gebiete der Naturwissenschaft*, I, 1876.

103. *Deutschlands Boden, sein geologischer Bau und dessen Einwirkungen auf das Leben der Menschen*, 2 Bde., Leipzig, 1854.

104. 501 pp., Breslau: F. Hirt, 1855.

105. Kutzen himself says in the *Vorwort* that he "leans on" Cotta; he cites the latter, for instance, on p. 466.

106. *Die Naturgeschichte des Volkes als Grundlage einer deutschen Sozialpolitik*, vol. 1 (11th ed., Stuttgart: Cotta, 1908): Land und Leute.

107. *Vide* the first Preface, written in 1853, to volume one, pp. VI-VII.

108. *Die Naturgeschichte*, etc., I, p. 42.

109. Ibid., Vorwort zur achten Auflage, 1883, p. X.

110. Die Naturgeschichte, etc., Vierter Band, "Wanderbuch," als zweiter Teil zu "Land und Leute." Vierte Aufl., 1903, p. 32.

111. G. P. Gooch, *History and Historians in the Nineteenth Century* (London & N. Y.; Longmans, Green & Co., 1913), p. 576.

112. Gooch, *ibid.*, p. 575.

113. For Riehl's view of milieu in a scheme of sciences, cf. *Die Naturgeschichte*, etc., I, pp. 40-2.

114. 164 pp., Meyers Volksbücher, Leipzig u. Wien: Bibliographisches Institut, *s.a.*—This essay forms the second chapter in Hans Meyer's *Das deutsche Volkstum* (2. Aufl., 1903), pp. 41-122.

115. Moderne Völkerkunde, p. 81, n.

116. 2. Aufl., 1905 (*Aus Natur und Geisteswelt*, 31. Bändchen, Leipzig: B. G. Teubner), 127 pp.—It has been translated into English under the title *Man and Earth* (London & N. Y., 1906. Reprinted 1914, 223 pp.) by A. S. "from the second amended German edition," in which are intercalated two chapters: Chapter V, on *The British Isles and Britons*, by the author; and Chapter VI, on *America and the Americans*, by the translator.—The first four chapters of a general nature—features of the globe, sea, steppes and deserts, in their influence on civilization, the influence of man on landscape—are followed by four chapters on *The British Isles and Britons, America and the Americans, Germany and the Germans, China and the Chinese*.

117. Vorlesungen, etc., delivered at Berlin in 1891 and 1897/8.

118. "… Es ist mehr unsere Aufgabe gewesen, in dem großen Getriebe der Siedlung und des Verkehrs der *allmählichen Entwicklung* nachzugehen, das steigende Maß der Überwindung von Widerständen durch den Menschen zu zeigen, die Kräfte zu untersuchen, welche in der Entwicklung wirksam sind,—als bei der großen Fülle des Tatsächlichen der heutigen Zeit zu verweilen." *Vorlesungen*, p. 351.

119. It will be noted that Herder is not mentioned here.

120. Ellen C. Semple, *Influences of Geographic Environment* (N. Y., 1911), p. V.

121. "In Germany the exponents of these theories of environmental influence were Cotta and Kohl, and later Peschel, Kirchhof, Bastian, and Gerland; but the greatest name of all is that of Fr. Ratzel, who has written the standard work on *Anthropogeographie*."—Haddon and Quiggin, *Hist. of Anthropology* (London, 1910), p. 152.—The first vol. of Ratzel's *Anthropogeographie* was published in 1882, 2nd ed. in 1899, the second vol. in 1897.

122. As further illustration, it might be instructive to compare here the chapter headings of Semple's *Influences of Geographic Environment*, which book was written "On the Basis of Ratzel's System of Anthropo-geography." They are as follows: I—Operation of Geographic Factors in History (1-31); II—Classes of Geographic Influences (22-50); III—Society and State in Relation to the Land (51-73); IV—Movements of Peoples in Their Geographical Significance (74-128); V—Geographical Location (129-67); VI—Geographical Area (168-203); VII—Geographical Boundaries (204-41); VIII—Coast Peoples (242-91); IX—Oceans and Enclosed Seas (292-317); X—Man's Relation to the Water (318-35); XI—The Anthropo-geography of Rivers (336-80); XII—Continents and Their Peninsulas (380-408); XIII—Island Peoples (409-72); XIV—Plains, Steppes and Deserts (473-523); XV—Mountain Barriers and Their Passes (524-56); XVI—Influences of a Mountain Environment (557-606); XVII—The Influences of Climate upon Man (607-37).

123. *Richthofen's Vorlesungen*, p. 13.

124. 1897; 2. Aufl. 1903.

125. "Diese die enge Erdgebundenheit in ihrer ganzen tiefgreifenden Bedeutung für das staatliche Leben erkannt und dargelegt zu haben, bleibt freilich für immer ein großes Verdienst der 'Politischen Geographie' ..."—O. Schlüter, "Die leitenden Gesichtspunkte d. Anthropogeogr.," *Arch. f. Sozialwiss.*, Bd. IV, p. 620.

126. Vide Richthofen, *l.c.*, p. 12.

127. 2 vols., München, 1893; see vol. 2, 2nd ed.: *Politische Geographie der Vereinigten Staaten von Amerika, unter besonderer Berücksichtigung der natürlichen Bedingungen u. wirtschaftlichen Verhältnisse* (763 pp.), esp. pp. 1–176.

128. London, 1896 (this is a translation of his *Völkerkunde*, 1887/8), cf. the opening pp. of vol. 1.

129. In Helmolt, *The History of the World* (N. Y., 1902), vol. 1, pp. 62–103, where Ratzel discusses in turn The Coherence of Countries, The Relation of Man to the Collective Life of the Earth, Races and States as Organisms, Historical Movement, Natural Regions, Climate and Location, Geographical Situation, Area, Population, The Water-Oceans, Seas, and Rivers, Conformation of the Earth's Surface.

130. London & N. Y.: Longmans, 1915.

131. See *The Nation*, N. Y., March 18, 1915, p. 310.

132. Paris, 1911, 420 pp.

133. Semple, *l.c.*, p. VI; cf. also Ratzel, *Anthropogeogr.*, I,$_2$ p. XII.

134. *Archiv f. Sozialwissenschaft*, Bd. IV (1906), pp. 581–630.

135. For Ratzel, cf. also Paul Barth, *Die Philosophie der Geschichte als Soziologie*, I (Leipzig: Reisland, 1897), pp. 227–30; Jean Brunhes, *La Géographie Humaine*, 2ᵉ éd. (Paris: Alcan, 1912), pp. 39–47.

136. Buckle, History of Civilization (1867), p. 32 n.

137. Robertson, *Buckle and his Critics* (London, 1895), p. 8 n.

138. 4. vols., 1822–3.

139. Flint, *l.c.*, pp. 577–9. See also p. 576.

140. *Vide supra* my note no. 84.

141. Flint, *l.c.*, p. 467.

142. *The History of Civilization from the Fall of the Roman Empire to the French Revolution* (4 vols., translated by Wm. Hazlitt, N. Y.: D. Appleton & Co., 1867—the lectures were delivered in the years 1828, 1829, and 1830), vol. 2, pp. 109 f.

143. "Gothein had attracted attention by a study of the civilisation of Southern Italy, which he had traversed on foot as Riehl had traversed the Palatinate."—Gooch, *l.c.*, p. 587.

144. "Voila pourquoi il Michelet va en Italie avant d'écrire son *Histoire Romaine*; il veut avoir l'impression, le contact du sol, du climat, du paysage."—Lanson, *Hist. de la Litt. Franç.* (1912), p. 1021 n.

145. Abry-Audic-Crouzet, *Littérature Française* (3ᵉ éd., Paris, 1916), p. 580.

146. Jules Simon, *Mignet, Michelet, Henri Martin* (Paris, 1890), p. 191.

147. Flint, *l.c.*, p. 540.

148. Philos. Erdk. als wissenschaftliche Darstellung der Erdverhältnisse u. des Menschenlebens nach ihrem inneren Zusammenhange, 2 vols., Braunschweig, 1845; the 2nd ed. appeared in 1868 under the title *Allgemeine Vergleichende Erdkunde*.—This book holds a high place in Ratzel's estimation: "Kapp, dessen Philos. Erdk. eine tiefgedachte, von überragendem philosophischem Standpunkte aus gewonnene Übersicht der Naturbedingtheit des Geschichtsverlaufes in den größten Zügen entrollt, ..."—Ratzel, *Anthropogeographie*, I^2, p. 34.

149. See Achelis, *l.c.*, pp. 76 f.

150. Brunhes, *l.c.*, p. 38 n.

151. Boston, 1849—It has been translated into English under the title *The Earth and man, or Physical geography in its relation to the history of mankind, Slightly abridged, etc.* (London: Parker, 1852), and into German as *Grundzüge der vergleichenden physikalischen Erdkunde in ihrer Beziehung zur Geschichte des Menschen* (1851).

152. (N. Y.: D. Appleton & Co., 1867—first published in 1857–61), vol. I, pp. 29–106: Influence exercised by physical laws over the organization of society and over the character of individuals.

153. Buckle and his Critics, London, 1895, 548 pp.

154. Camille Vallaux, *Géographie Sociale* (Paris, 1911), p. 23.

155. Vide supra, p. 46 f.

156. Anthropogeographie, I^2, p. 87.

157. The German original appeared in 1857–67, and the English translation by A. W. Ward in 1868–73.

158. New York: Scribner, vol. I (1871), pp. 9–46; cf. esp. pp. 9–25, 34, 37.

159. Boden und Klima von Athen. Rede in der öffentlichen Sitzung der Kgl. Akademie der Wissenschaften am Leibniztage 5. Juli 1877 (15 pp.).

160. For the same, cf. also H. Koester "Über den Einfluß landschaftlicher Verhältnisse auf die Entwicklung des attischen Volkscharakters" (Progr., Saarbrücken, 1898).

161. E.g. by Ratzel, jointly with Curtius' account thereof. Cf. *Anthropogeogr.*, I², p. 37.

162. In 12 vols., vol. II (London: John Murray, 1869), Part II, ch. I, pp. 213-37.

163. Political effects of locality: strengthened defense; difficulty of attack; politically disunited; indefinite multiplication of self-governing cities.

164. Intellectual effects of locality: the geographical position made them mountaineers and mariners; variety of experience; each petty community possessed an individual life, yet sympathized with the remainder; commerce with a great diversity of half-country-men; Grecian festivals; Homer dependent upon the conditions of his age.

165. Oxford, Clarendon Press (1911, 454 pp.), pp. 13-64. "It is now generally admitted that neither an individual nor a nation can be properly understood without a knowledge of their surroundings and means of support—in other words, of their geographical and economic conditions."—*Ibid.*, Preface, p. 5.

166. Zimmern refers in this book—*e.g.* p. 18, 41, 43, *et al.*—to the writings of Myres: "Greek Lands and the Greek People," "Herodotus and Anthropology" (in "Anthropology and the Classics"), and "The Geographical Aspect of Greek Colonization" (in *Proceedings of the Classical Association*, vol. VIII—1911).—Cf. also H. Dondorff, *Das hellenische Land als Schauplatz der althellenischen Geschichte, in Sammlung gemeinverständlicher wissenschaftlicher Vorträge, begründet von Virchow u. Holtzendorf*, 1889, Neue Folge, Serie 3, Heft 72.

167. Revised ed., in 2 vols. (N. Y.: Harper & Brothers, 1876). The Preface of the first ed. is dated 1861.

168. Heinrich Boehmer, *Geschichte der Entwicklung der naturwissenschaftlichen Weltanschauung in Deutschland* (Gotha, 1872, 232 pp.), p. 195: "... Hersche Ideen waren leitend für den Aufbau der Geschichte."

169. Leipzig, 1878-86.

170. Cited by Achelis, *l.c.*, p. 84.

171. Ibid., pp. 85 f.

172. Ibid., p. 86.

173. "... Indessen darf man nicht vergessen, daß die allgemeine Gestalt der Kontinente und der Meere und aller besonderer Züge der Erde in der Geschichte der Menschheit einen wesentlich wechselnden Wert besitzen, je nach dem Stande der Kultur, auf welchem die Nationen angelangt sind ..."—*Ibid.*

174. Ibid., p. 87.

175. Paris, 1886.

176. Vide P. Barth, *Die Philosophie der Geschichte als Soziologie* (Leipzig, 1897), p. 230.

177. See Barth, *l.c.*, pp. 231 f.

178. Ibid., p. 233.—Mougeoulle makes the milieu account for the great men in history, the great popular epics, social and historical life in general; the tendencies of the three historical schools—

German, French, and English—are connected with the differences in the milieus of their respective countries.—Cf. *ibid.*, pp. 230–2.

179. Avec une Préface de M. Élisée Réclus (Paris: Hachette, 1889, 369 pp.), pp. 53–71.

180. Ibid., p. 156; 130.

181. Ibid., p. 154; 157 f.

182. Ibid., p. 278; 190 ff.; 188; 135.—But why does he confine himself to these four countries?

183. Ibid., p. 185; 364. For a general statement on the significance of rivers, cf. *ibid.*, pp. 188–90. The particular nature of the rivers of the "territoire des civilisations fluviales" imposed on the inhabitants the yoke of despotism.—*Ibid.*, p. 161.

184. Ibid., pp. 364 f.

185. Ibid., p. 364.

186. Ibid., e.g., p. 128; 224–27.

187. His general theory is stated on pp. 39–42, 53–71, 79 f., 89, 99 f., 102–60. Chapter 7, pp. 161–90, is a general discussion of the geographical environment of the "Civilisations Fluviales," followed successively by a detailed treatment of "Le Nil" (ch. 8, pp. 191–234), of "Le Tigre et L'Euphrate" (ch. 9, pp. 235–78), of "L'Indus et Le Gange" (ch. 10, pp. 279–319), of "Le Hoang-Ho et Le Yangtse-Kiang" (ch. 11, pp. 320–66).

188. W. D. Babington, *Fallacies of Race Theories as Applied to National Characteristics* (Longmans, Green & Co., 1895).

189. N. Y., Scribner, 1893, 290 pp.

190. For the rôle of the physical milieu in American history, cf. also: Justin Winsor, *The Mississippi Basin, The Struggle in America between England and France: 1697–1763* (Boston & N. Y., 1898) influence of geography over history during colonization and settlement; Frederick Jackson Turner, *Rise of the New West*: 1819–1829 (N. Y. & London: Harper & Brothers, 1906) vol. 14 of *The American Nation, A History*, ed. by A. B. Hart, in 27 vols. In the Author's Preface, p. XVII, Turner remarks: "In the present volume I have kept before myself the importance of regarding American development as the outcome of economic and social as well as political forces." And, he should have added, of geographical environment. *Vide* especially the first half of his book for the working out of his milieu idea; James Bryce, *The American Commonwealth*, (2 vols., new ed., completely revised, N. Y.: Macmillan, 1910–11) see vol. 2, ch. 91 (pp. 449–68), "The home of the nation," for a statement of the influence of physical conditions on American history; E. C. Semple, *American History and Its Geographic Conditions* (Boston & N. Y.: Houghton, Mifflin & Co., 1903, 435 pp.) regarded, I believe, as one of the best treatises on the subject; A. P. Brigham, *Geographic Influences in American History* (Boston: Ginn, 1903, 355 pp.) a concrete essay; has much physiography; includes present conditions; A. M. Simons, *Social Forces in American History* (N. Y.: Macmillan, 1914, 325 pp.) a discussion of the effect of the industrial and economic environment on social institutions in America; perhaps it may be added here that some American universities offer a course on the relation of geography to American history.

191. See Ripley, "Geography and Sociology" (1895), p. 637; and Ripley, *The Races of Europe* (1899), pp. 4 ff.; for titles of their writings on this subject, cf. *ibid.*, pp. 4–6 nn., and "Geogr. and Soc.," pp. 654 f.

192. 8 vols., N. Y., Dodd, Mead & Co., 1902–7.

193. See Bryce's article in Helmolt's *Hist. of the World*, vol. 1, p. XL.

194. "Anderseits wieder hat ja Helmolt in seinem geschichtlichen Sammelwerke im Geiste Ratzels den Versuch gemacht, ein Gesamtgeschichtsbild auf geographischer Grundlage aufzubauen, so daß kein Teil der Ökumene aus der Weltgeschichte ausgeschlossen bleibt."—L. Gumplowicz, Der *Rassenkampf* (2 .... Aufl., 1909), p. 403 (Anhang).

195. "... die bisherigen Weltgeschichten waren gar keine Geschichte der Welt oder auch nur unserer Welt, sondern einzig eine solche der Kulturnationen. Mit dieser Gepflogenheit hat Helmolts Werk in ebenso glücklicher wie origineller Weise gebrochen, indem es zum ersten Male die Länder- und Völkerkunde in den Dienst der Weltgeschichtsdarstellung hineinzog."—From a review of the first ed. of *Helmolts Weltgeschichte* (1899) in the "Braunschweigische Landeßeitung" (February 4, 1908), quoted in the prospectus of the second German edition.

196. *History and Historians in the Nineteenth Century* (London, 1913).

197. Second ed., Oxford, The Clarendon Press, 1903, 288 pp.

198. George, *l.c.*, p. V (Preface).

199. *Ibid.*, pp. 111 f.—George cites no authorities or sources; he has no bibliography; he does not quote a single book in his discussion; he has no *Auseinandersetzung* with his predecessors in the field; and finally, he gives no clue as to the origin of his data.—Chaps. 1-8 (pp. 1-110) are the general part of the book; chaps. 9-20 (pp. 111-282) deal with: The Outlines of Europe, The British Islands, France, The Spanish Peninsula, Italy, The Alpine Passes, Switzerland, The Rhineland, The Baltic Region, The Danube Basin, Theatres of European War, The Mediterranean Basin.

200. A. W. Small, *General Sociology* (Chicago, 1905), p. 53.

201. The distinguished Italian historian is the son-in-law of the late eminent Italian criminologist Cesare Lombroso.

202. Vide Jean Brunhes, *La Géographie Humaine* (2ᵉ éd., Paris, 1912), p. 721.—For references to historical works dealing with history on a geographical basis, cf. *ibid.* (1ᵉ éd., Paris, 1910), ch. X, 1: L'esprit géographique dans les sciences économiques, sociales et historiques (pp. 739 ff., esp. 774 ff. Michelet, Vidal de la Blache, Th. Reinach, A. Leroy-Beaulieu, C. Jullian, A. Harnack, H. F. Helmolt, G. Ferrero, E. C. Semple, Erwin Hanslick, & o.).

203. Die geographischen Grundlagen der österreichisch-ungarischen Monarchie u. ihrer Außenpolitik (Leipzig u. Berlin: B. G. Teubner, 1915).

204. See the review of Sieger's book by Edwin Rollett in the *Österreichische Rundschau*, Bd. 43, H. 4 (15. Mai 1915), pp. 188 f.

205. Boston & N. Y., Houghton, Mifflin & Co., 1907.

206. Cf. esp. ch. 18 (pp. 359–85) for a summary of conclusions.

207. Vide e.g. James Harvey Robinson's *The New History, Essays Illustrating the Modern Historical Outlook* (N. Y.: Macmillan, 1912), for references to the theory of milieu, cf. esp. p. 64, 73, 76 f., 92 f., 97 f., 124–6, 144, 145 f., 247, 253–7, and ch. 3 (pp. 70 ff.): The new allies of history. Or take for choice the title of a recent book by Charles A. Beard: *An Economic Interpretation of American Politics* (Macmillan, 1916), to be further persuaded of the attention bestowed by historians on the milieu. Or, see works by Seligman and J. T. Shotwell.

208. Vide C. Vallaux, *Géographie Sociale, Le Sol et L'État* (Paris, 1911), p. 23.—Such economists as Blanqui, Bastiat, and J.—B. Say, brought to light the geographical bases of the material life of societies. The sociologists themselves, "bien que leur science soit jeune, n'ont pas toujours oublié le cadre naturel et la position terrestre des agrégats qu'ils étudient. Par tous ces chercheurs de tendances diverses, la géographie humaine et la géographie politique ont progressé tout autant que par les efforts des géographes proprement dits."—*Ibid.*

209. E. Bernheim, *Lehrbuch der historischen Methode* (5. u. 6. Aufl., Leipzig, 1908), p. 316; 636.—Cf. also E. Fr. Th. Lindner, *Geschichtsphilosophie, das Wesen der geschichtlichen Entwicklung* (2. erweiterte u. umgearb. Aufl., Stuttg. u. Berlin: Cotta, 1904, 241 pp.), 2. Abschnitt (pp. 23–34): Die Veränderung, but more esp. 10. Abschnitt (pp. 217–41): Die Ursachen u. die Weise der Entwicklung.

210. For orientation and literature on views opposing the naturalistic interpretation of history, cf. L. Stein, *Philosophische Strömungen der Gegenwart* (Stuttgart, Verl. v. F. Enke, 1908), pp. 430 ff.

211. See G. P. Gooch, *History and Historians in the Nineteenth Century* (London & N. Y.: Longmans, Green & Co., 1913), p. 573; see ch. 28 (pp. 573–94): "The History of Civilisation;" also *The Cambridge Modern History* ed. by A. W. Ward and others, Cambridge: The Univ. Press, 1910, vol. 12: *The Latest Age*, ch. 26 (pp. 816 ff.: "The Growth of Historical Science" by G. P. Gooch).

212. Economic Geography (N. Y.: Macmillan, s.a.—1915?—; not earlier than 1910, for statistics for that year are given in the text; 560 pp.), p. 1.

213. "Since his Buckle's time much more has been done, not only in studying, as Buckle himself did, the immediate influence of climate and soil, but also in explaining the allied field of the effect of the fauna and the flora on social development. The subject of the domestication of animals, for instance, and its profound effect on human progress has not only been investigated by a number of recent students especially E. Hahn, *Die Haustiere u. ihre Beziehung zur Wirtschaft des Menschen*, 1896, but has been made the very basis of the explanation of early American civilization by one of the most brilliant and most learned of recent historians Payne, *History of the*

New World called America; esp. vol. 1, bk. II. A Russian scholar has shown in detail the connection between the great rivers and the progress of humanity, and the whole modern study of economic geography is but an expansion on broader lines of the same idea."—Edwin R. A. Seligman, *The Economic Interpretation of History* (N. Y.: The Columbia Univ. Press, 1902, 166 pp.), pp. 13 f.

214. See Wm. Morris Davis, *Geographical Essays*, ed. by D. W. Johnson (Ginn & Co.: Boston, *s.a.*, copyright 1909), esp. the first two essays: "An inductive study of the content of geography" (1906), pp. 3–22, and "The progress of geography in the schools" (1902), pp. 23–69.

215. In an address delivered at the dedication of Julius Rosenwald Hall, printed in *The University of Chicago Magazine* (vol. VII, No. 6—April, 1915—, pp. 175-8) under the title "Some Matters of History." See p. 177.

216. Felix Lampe, in *Große Geographen* (Leipzig, 1915), has a rather brief chapter (pp. 281–7) on "Die wissenschaftliche Geographie der Gegenwart."

217. See the Introductory Essay by the Right Hon. now Viscount James Bryce in Helmolt's *Hist. of the World*, vol. 1, pp. I-LX, esp. pp. XXV-XLI.

218. A. J. Herbertson and F. D. Herbertson, *Man and his Work, an Introduction to Human Geography* (London: Black, 1909, 132 pp.), p. 6.

219. N. Y., G. P. Putnam's Sons, 1908, 363 pp.

220. "In the chapters on the life of man in the different zones, I have made liberal use of Ratzel's *Anthropogeographie* (2d ed., Stuttgart, 1899)."—Ward, *op. cit.*, p. VI.

221. Ward, *op. cit.*, p. V.

222. N. Y. and London, 1911. See ch. 4, pp. 94–129.

223. Paris, 1911, 420 pp.

224. Vide supra, p. 27.

225. "Die soziale Geographie, hauptsächlich von Bastian und Ratzel tiefer begründet, wird gegenwärtig immer sorgsamer ausgebaut und hat namentlich in dem Wiener Erwin Hanslick einen eifrigen Förderer, der auf die Ermittlung von geographischen Kulturgrenzen ausgeht. In andrer Weise nimmt von ihr Willy Hellpach seinen Ausgang, der Geographie, Psychologie und Soziologie zu einem neuen Gebiet zu vereinigen sucht."—Rudolf Goldscheid, "Soziologie" in *Das Jahr 1913, Ein Gesamtbild der Kulturentwicklung*, herausgegeben von D. Sarason (Leipzig und Berlin: B. G. Teubner, 1913), p. 432.

226. Leipzig, W. Engelmann, 1911, 368 pp.—"Hier in Hellpach's book wird alles zusammengefaßt, was über den Einfluß von 'Wetter, Klima und Landschaft' auf das Seelenleben bekannt ist."—Otto Schlüter, "Anthropogeographie" in *Das Jahr 1913*, etc., p. 401.

227. See Hellpach, *op. cit.*, p. 4.—Chiefly with those of the atmosphere; he devotes nine pages (98–107) to the telluric elements of the weather, and 87 pages (230–317) to the third main part of the book: "Landschaft und Seelenleben." For soil as a co-factor, cf. also the ch. "Klimawechsel" in Part II (pp. 118–38). Hellpach defines Landschaft (p. 230) as follows: "Unter Landschaft verstehen wir den *sinnlichen* Gesamteindruck, der von einem Stück der Oberfläche und dem dazu gehörigen Abschnitt des Himmelsgewölbes in uns erweckt wird. ... das *sicht*bare Landschaftsbild bildet unter allen Umständen den Kern dessen, was wir Landschaft nennen ... And he adds that for an investigation of the effect of Landscape upon the human soul sind die nicht-optischen sinnlichen Eigenschaften der Landschaft von unentbehrlicher Bedeutung: Töne und Geräusche, Düfte und Gerüche und eine höchst verwickelte Summe von Affizierungen der Berührungs-,

Temperatur-, ja zuweilen der Schmerzempfindlichkeit erst bilden mit Farben und Formen zusammen das natürliche Ganze, das wir in seelischen Wirkungen als *Landschaft* erleben."

228. Vide, *e.g.*, p. 8.

229. Hellpach himself testifies (p. 318) that his book is a "Sammlung der Tatsachen." Cf. also Schlüter's opinion cited above in note no. 226.

230. Manifestly, this is to be understood as a virtue in Hellpach, and not as a fault, since this conviction is gained only by dint of Hellpach's clear delimitation of the scope of his work; it constitutes one of the results of his own labor.

231. See Schlüter's art. in *Das Jahr 1913*, p. 402.

232. Paris, 1910; 2nd ed. 1912.

233. For a statement of principles (theoretical exposition), cf. the first two chaps. (pp. 1–92); for a summary, cf. ch. X, section 2 (pp. 780–9): "Le facteur psychologique dans les phénomènes naturels et l'activité humaine," and section 3 (pp. 790–807): "L'adaptation humaine aux conditions géographiques." In the preface to the second ed., there are quoted seven pages from a review of the first ed. of Brunhes' work by Paul Mantoux, wherein the scope, content, and import of the first ed. are succinctly summarized.

234. N. Y., 1911, 637 pp.

235. Vide Wm. J. Thomas, *Source Book for Social Origins* (Chicago and London, 1909), p. 138 (Bibliogr. to Part I).—Without fear of contradiction, it may be said that the best two recent treatises on human geography are those by Brunhes and Semple.—For a brief concrete anthropo-geographical sketch, besides the works previously cited, cf. also W. Ule, *Grundriß der Allgemeinen Erdkunde* (2. verm. Aufl., Leipzig: S. Hirzel, 1915, 487 pp.), pp. 361 ff. See also the brief résumé in G. Schmoller's *Grundr. d. Allgem. Volkswirtschaftslehre* (Leipzig, 1901), pp. 144 ff.

236. "Unverkennbar ist es, daß die Naturgewalten in ihren bedingenden Einflüssen auf das Persönliche der Völkerentwicklung immer mehr und mehr zurückweichen mußten, in demselben Maße wie diese vorwärts schritten. Sie übten im Anfange der Menschengeschichte als Naturimpulse über die ersten Entwicklungen in der Wiege der Menschheit einen sehr entscheidenden Einfluß aus, dessen Differenzen wir vielleicht noch in dem Naturschlage der verschiedenen Menschenrassen oder ihrer physisch verschiedenen Völkergruppen aus einer gänzlich unbekannten Zeit wahrzunehmen vermochten. Aber dieser Einfluß mußte abnehmen, ... Die zivilisierte Menschheit entwindet sich nach und nach, ebenso wie der einzelne Mensch, den unmittelbar bedingenden Fesseln der Natur und ihres Wohnortes. Die Einflüsse derselben Naturverhältnisse und derselben tellurischen Weltstellungen der erfüllten Räume bleiben sich also nicht durch alle Zeiten gleich." Ritter, *l.c.*; see Achelis, *op. cit.*, p. 74 *et seq.*

237. "Man ist in Nachfolge C. Ritters vielfach geneigt, anzunehmen, daß die Natureinflüsse sich mit zunehmender Kultur immer weniger geltend machen."—E. Bernheim, *Lehrb. d. hist. Methode* (Leipzig, 1908), p. 642.

238. Theo. Waitz, *Anthropologie der Naturvölker*, I (Leipzig, 1859), p. 341; see Achelis, *op. cit.*, p. 185.

239. "Die Einteilung der Menschheit war nur geographisch-historisch möglich. Denn der Mensch steht in fester Abhängigkeit, in engstem Verbande zu der Natur, aus und an welcher er sich entwickelt hat, zur Natur der Erde, welcher letzteren kleiner, aber integrierender Teil er ist. Auch seine Entwicklung ist noch im Steigen, aber nur im Bereiche seines inneren, geistigen Lebens ... je höher der Mensch steigt, um so mehr macht er sich von dem zwingenden Einfluß der Erde frei; und wenn er demselben auch nie ganz entgehen wird, da er Nahrung braucht, von der Schwere sich nicht loslösen kann, so ist dennoch diese immer wachsende Freiheit ... eine stärkende ... Aussicht für die Zukunft ..."—*Anthropologische Beiträge*, 1. Bd. (Halle, 1875), p. 423; see Achelis, *op. cit.*, p. 227.

240. *Principles of Sociology*, I, sec. 21.

241. Vide Ripley, "Geography and Sociology," p. 649.

242. *Contributions to the Theory of Natural Selection*, p. 319; cited by E. B. Tylor in the article "Anthropology," *Ency. Brit.* (11th ed.), vol. 2, p. 114.

243. Réclus, *op. cit.*, (1879); quoted by Achelis, *l.c.*, pp. 86 f.

244. "... je crois, que la civilisation dans son premier stade dépend bien plus du milieu physique et tellurique, qu'aux époques suivantes."—Aug. Matteuzzi, *Les Facteurs de l'Évolution des Peuples* (Paris, 1900), p. 29. "... Tout ceci nous amène à affirmer ce fait, que les premières civilisations, dans des milieux favorables, eurent une relation étroite avec la culture du sol; et que dans un développement ultérieur, ce rapport se relâcha ..."—*Ibid.*, p. 25. For best summaries of immense material collected on the relation of primitive human life to environment, see the five papers in the *Smithsonian Report* for 1895: "Relation of Primitive Peoples to Environment" by J. W. Powell (pp. 625 ff.); "Influence of Environment upon Human Industries or Arts" by O. T. Mason (pp. 639 ff.); "The Japanese Nation—A Typical Product of Environment" by G. G. Hubbard (pp. 667 ff.); "The Tusayan Ritual: A Study of the Influence of Environment on Aboriginal Cults" by J. W. Fewkes (pp. 683 ff.); and, probably the best of the five, "The Relation of Institutions to Environment" by the eminent ethnologist W. J. McGee (pp. 701 ff.).

245. *Anthropogeogr.*, I$_2$: "Der Mensch und die Umwelt" (pp. 41–65).

246. "Geogr. and Sociol.," p. 650.

247. See his presidential address on the Origin of Man before the Section of Anthropology (*Report of the British Association for the Advancement of Science, 1912*; London, 1913), p. 576.

248. The Positive Philosophy of Aug. Comte, Freely Translated and Condensed by Harriet Martineau *(In 2 vols., 3rd ed., London, 1893— the original appeared from 1830–42), vol. 2, p. 96.*

249. Aug. Comte's Positive Philosophie im Außug von I. Rig, Übersetzt von Kirchmann (2 Bde, Heidelberg, 1883), S. 94 ff.; Achelis, *op. cit.*, p. 130.

250. A System of Logic (New Impression; London: Longmans, Green & Co., 1911—first published in 1843), p. 572.

251. A. Schäffle, *Bau und Leben des sozialen Körpers*, Tübingen, 1875, 2. Aufl., 1881; Achelis, *op. cit.*, p. 161.

252. "Post's general attitude is best seen in his 'Introduction to the Study of Ethnological Jurisprudence,' which was published in 1886, and in his 'African Jurisprudence' of 1887."—John L. Myres, "The Influence of Anthropology on the Course of Political Science" (Presidential address to the Anthropological Section of the British Assoc. for the Advancement of Science), *Report Brit. Assoc., 1909* (London, 1910), p. 613.

253. Myres, *ibid.*, pp. 613 f.

254. See Rob. DeC. Ward, *op. cit.*, p. 231.

255. See the 4th ch. of his *Géographie Sociale* (Paris, 1911): "Agents et Caractères Physiques Considérés Isolément" (pp. 92–144).

256. "... as political and legal institutions are indissolubly bound up with social and religious, it follows inevitably that the political and legal institutions of a race cradled in Northern Europe are exceedingly ill adapted for the children of the equator. Accordingly in any wise administration of these regions it must be a primary object to study the native institutions, to modify ... them ..., but never to seek to eradicate and supplant them. Any attempt to do so will be but vain, for these institutions are as much part of the land as are its climate, its soil, its fauna, and its flora. 'Naturam expellas furca, tamen usque recurret.'"—The Application of Zoological Laws to Man, in *Rep. Brit. Assoc, f. the Adv. of Sci., 1908* (London, 1909), p. 843.

257. Rob. DeC. Ward, *op. cit.*, pp. 310 *et seq.*

258. Vide pp. 141–75 in *Der Weltkrieg im Unterricht, Vorschläge u. Anregungen*, etc. (Gotha: F. A. Perthes), esp. pp 163–5; he also discusses other phases of the relation between physical environment and the present war.

259. I: *Deutsche Rundschau*, April, 1915, pp. 78–91, and II (Schluß): *ibid.*, May, 1915, pp. 207–17.

260. In *Monatshefte für den Naturwissenschaftlichen Unterricht*, 1. Kriegsheft von Bastian Schmid (Leipzig: B. G. Teubner, 1915).

261. Cf. Gooch, *op. cit.*, pp. 585 *et seq.*

262. See his Introduction to Dexter's *Weather Influences* (N. Y., 1904), p. XXIV.

263. Les Facteurs de L'Évolution des Peuples (Paris, 1900), p. 25, 29, 27.—"C'est dans l'intensité de l'effort dirigé par les groupes sociaux contre les résistances du milieu, que réside la première impulsion vers la civilisation."—*Ibid.*, p. 27.

264. But he adds, "... no disturbing causes, acting on social development, could do more than to affect its rate of progress. This is true of the operation of influences from the inorganic world, as of all others. In our view of biology we saw that the human being cannot be modified indefinitely by exterior circumstances; that such modifications can affect only the degrees of phenomena, without at all changing their nature; and again, that when the disturbing influences exceed their general limits, the organism is no longer modified, but destroyed."—*The Positive Philosophy of Aug. Comte*, tr. by Harriet Martineau, vol. 2, p. 98; 97.

265. See Ripley, *Races of Europe* (1899), p. 11; cf. the references given there, and in the note on the same page.—Cf. also Ellsworth Huntington's *Palestine and its Transformation* (1910), and his suggestive articles on "Changes of Climate and History" (in *The American Historical Review* for January, 1913, vol. 18, pp. 213–32) for references to other writings on the subject by the same author,— and by A. T. Olmstead—cf. p. 214 n.; on "Climate and Civilization" (in *Harper's Magazine* for February, 1915, vol. 130, pp. 367–73); on "Is Civilization Determined by Climate?" (*ibid.* May, 1915, pp. 943–51); a new book of his, entitled *Civilization and Climate* (333 pp.), is announced for publication by the Yale Univ. Press.

266. Rob. DeC. Ward, *op. cit.*, pp. 280 *et seq.*

267. "... cetera Mattiaci similes Batavis, nisi quod ipso adhuc terrae suae solo et caelo acrius animantur."—F. Ritter, *P. C. Taciti Opera* (1864), p. 643. In *Römische Prosaiker in neuen Übersetzungen* (hg. v. C. N. von Osiander und G. Schwab, 51. Bändchen, Stuttg., 1852, S. 123) this is rendered as follows: "Im ganzen gleichen sie die Mattiaker den Batavern, nur daß Boden und Klima ihres Landes sie noch kriegerischer macht."

268. Cesare Lombroso, *Crime, Its Causes and Remedies* (Boston, 1911), pp. 3 f.

269. Rob. DeC. Ward, *op. cit.*, p. 282.

270. Vide Flint, *l.c.*, pp. 582 *et seq.*

271. Haddon & Quiggin, *Hist. of Anthropology* (London, 1910), pp. 84 *et seq.*

272. Cesare Lombroso, *Crime*, etc., p. 2.

273. N. S. Shaler, Nature and Man in America (N. Y., 1893), p. 288.

274. In *Abhandlungen der Königl. Preuss. Akademie der Wissenschaften, Phil.-hist. Classe*, 1912, p. 13: "In einer Wendung, die an Distinktionen Schleiermachers erinnert, hat er Dilthey in seiner letzten größeren Arbeit erklärt, daß unser wissenschaftliches Denken von zwei großen Tendenzen beherrscht sei. Der Mensch finde sich auf der einen Seite bestimmt von der physischen Welt, in der die seelischen Vorgänge nur wie Interpolationen erscheinen. The other is: das Leben (life), das Erlebnis (experience)."

275. Ridgeway, *l.c.*, p. 843.

276. Rob. DeC. Ward, *op. cit.*, pp. 258 *et seq.*—For the effect of physical environment on the Jews in Palestine, cf. Friedrich Otto Hertz, *Rasse und Kultur* (Leipzig, 1915, 421 pp.), pp. 162 ff.; and "Soziale Grundlagen des Monotheismus u. Polytheismus" (pp. 170 ff.) and the literature there cited. Cf. also *ibid.*, "Natürliche u. Soziale Grundlagen der indischen Entwicklung" (pp. 198 ff.).

277. Rob. DeC. Ward, *op. cit.*, pp. 309 *et seq.*

278. Vide his *Weather Influences, An Empirical Study of the Mental and Physiological Effects of Definite Meteorological Conditions*, with Introduction by Cleveland Abbe (N. Y.: Macmillan, 1904, 277 pp.).

279. I saw somewhere that exception had been taken to his results, but I failed at the time to make a note thereof and have been unable to find the passage again.

280. Ibid., p. 266; 269; 272 f.—The fifth and last is not cited here.

281. Ward, *op. cit.*, p. 310; 335, where ref. is also made to F. A. Cook's article on "Some Physiological Effects of Arctic Cold, Darkness and Light" (*MED. REC.*, June 12, 1897, pp. 833–36).

282. London and N. Y., 1892.

283. Ibid., p. 90.

284. Ibid., pp. 113–5.

285. "Diese Priorität (der erste Versuch überhaupt, die Einflüsse des naturalen Milieus auf die Psyche darzustellen) gebührt, nach mancherlei Vorläufern minder geschlossenen Charakters (z. B. *Quételet,* Sur l'homme etc. 1835, Bd. 2, Kap. 3, Abschn. 2–3, Influence du climat et des saisons sur le penchant au crime) ohne Zweifel *Lombroso,* aus dessen 1878 erschienenem Buche 'Pensiero e meteore' Extracte auch in seine andern Publikationen, namentlich in 'Genio e follia,' übergegangen sind."—Hellpach, *Die Geopsychischen Erscheinungen* (Leipzig, 1911), p. 336.

286. Criminal Man, According to the Classification of Cesare Lombroso Briefly Summarized by his Daughter Gina Lombroso Ferrero ("The Science Series"; N. Y. and London: G. P. Putnam's Sons, 1911, 322 pp.), p. 145.—Lombroso's *L'Uomo di genio* appeared in 1888, *L'Uomo delinquente* in 1889, and *La Donna delinquente* in 1893.

287. Criminal Man, p. 145.

288. Tr. by H. P. Horton, "The Modern Criminal Science Series," Boston: Little, Brown and Co., 1911, 471 pp.

289. "It is brought out in Guerry's statistics that the crime of rape occurs in England and France oftenest in the hot months; and Curcio has observed the same thing in Italy....

"In England, according to Guerry, and in Italy, according to Curcio, the maximum number of murders falls in the hottest months....

"Poisoning also, according to Guerry, occurs oftenest in May. The same phenomenon is to be observed in the case of Rebellions. In studying (as I have in my 'Political Crime') the 836 uprisings that took place in the whole world in the period between 1791 and 1880, one finds that in Asia and Africa the greatest number falls in July. In Europe and America the greater prevalence of rebellions in the hot months could not be more clearly marked. In Europe the maximum proved to be in July in this connection one might also point to the beginning of the present European war which falls in the midsummer of 1914, and in South America in January, which are respectively the two hottest months. The minimum falls in Europe in December and January, and in South America in May and June, which again correspond in temperature.

"If now we pass from the whole of Europe to the particular countries, we still find the greatest number of uprisings in the hot months....

"Benoiston de Chateauneuf points out that duels in the army are more frequent in the summer.

"I have proved that the same influence manifests itself in the case of men of genius ('Man of Genius,' Part I.).

"Ferri, in his 'Crime in its Relation to Temperature,' has proved from a study of the French criminal statistics from 1825 to 1878 that one can deduce an almost complete parallelism between heat and criminality, not only for the different months, but also for years of different degrees of heat. The influence of the temperature on crime from 1825 to 1848 appears to be very pronounced and constant, and is often even greater than that exercised by agricultural production. Since 1848, notwithstanding the more serious agricultural and political disturbances, the coincidence between temperature and criminality becomes from time to time plainly apparent, especially in the case of homicide and murder....

"The connection comes out much more plainly, however, in the statistics of rape and offenses against chastity, which follow to an even greater degree the annual variations in temperature....

"As regards crimes against property there is a marked increase in the winter (theft and forgery being the most abundant in January), while the other seasons differ little from one another...."—

Lombroso, *Crime, Its Causes and Remedies*, pp. 4–8. "Superintendents of prisons have generally observed that the inmates are more excited when storms are approaching and during the first quarter of the moon...."—*Ibid.*, p. 12.

290. *Ibid.*, p. 13.—"In studying the distribution of simple and aggravated homicides in Europe, we find the highest figures in Italy and the other southern countries, and the lowest in the more northerly regions, England, Denmark, Germany. The same can be said of political uprisings in all Europe. We see, in fact, that the number of crimes increases as we go from north to south, and in the same measure as the heat increases."—*Ibid.*, p. 14.

291. This follows Laing. See Robertson, *Buckle and his Critics* (London, 1895), p. 553.—Cf. also C. M. Gießler's article, "Über den Einfluß von Wärme und Kälte auf das seelische Funktionieren des Menschen," in *Vierteljahrsschrift für wissenschaftliche Philosophie u. Soziologie*, 1902, pp. 319–38. Gießler refers (p. 334) to Oppenheimer "Über den Einfluß des Klimas auf den Menschen" (Berlin, 1867). *Vide* also E. Huntington's article on "Work and Weather," *Harper's Magazine*, vol. 130 (January, 1915), pp. 233–44.

292. *Rep. Brit. Assoc., 1908* (London, 1909), p. 844.

293. On the use of alcohol in its relation to the northern climate, cf. also Auguste Matteuzzi, *Les Facteurs de L'Évolution des Peuples* (Paris, 1900), pp. 329 *et seq.*

294. Some of these are to be discussed in a subsequent paper.

www.ingramcontent.com/pod-product-compliance
Lightning Source LLC
Chambersburg PA
CBHW020450220526

45464CB00002B/940